From Research to Revolution

Scientific, Business, and Legal Perspectives on the New Biotechnology

Edited by:
Robert A. Bohrer

California Western School of Law

Proceedings of the First Annual San Diego
Biotechnology Conference

Sponsored by:
California Western School of Law
University of California at San Diego
The San Diego County Bar Association—
Section on High Technology and Computer Law

FRED B. ROTHMAN
Littleton, Colorado 80127
1987

Muirhead Library
Michigan Christian College
Rochester, Michigan

Library of Congress Cataloging-in-Publication Data

San Diego Biotechnology Conference (1st : 1985 : La
   Jolla, San Diego, Calif.
      From research to revolution.

   Bibliography: p.
   1. Biotechnology—Congresses.   2. Biotechnology
industries—Congresses.   I. Bohrer, Robert A.
II. California Western School of Law.   III. University
of California, San Diego.   IV. San Diego County Bar
Association. Section on High Technology and Computer
Law.   V. Title.
TP248.14.S26   1985   338.4'7660'6   87-9634
ISBN 0-8377-0355-7

©1987 by Robert A. Bohrer
All rights reserved.

Printed in the United States of America

TABLE OF CONTENTS

Foreword   vii

Acknowledgments   xi

Part One - OVERVIEW   1

   Chapter One - Dr. Ramon Pinon

   RECOMBINANT DNA: CONTROVERSY
   AND PROMISE: A SCIENTIST'S
   OVERVIEW   3

   Chapter Two - Vincent Frank, Esq.

   A GENERAL INTRODUCTION TO THE
   NEW BIOTECHNOLOGY BUSINESS   13

   Chapter Three - William L. Respess, Esq.

   BIOTECHNOLOGY: THE LAWYER'S ROLE   21

Comment – Robert A. Bohrer

LAWYERS, BUSINESS PERSONS AND
SCIENTISTS: BIOTECHNOLOGY
PROBLEM SOLVERS..33

Questions and Answers..35

Part Two – CURRENT ISSUES 39

Chapter Four – Dr. Ted Friedmann

HUMAN SOMATIC CELL GENE THERAPY.................41

Chapter Five – Ray McKewon

THE BUSINESS PERSON AND THE
UNIVERSITY...51

Chapter Six – Ned Israelsen, Esq.

CURRENT ISSUES IN PROPRIETARY
RIGHTS TO BIOTECHNOLOGY.................................59

Chapter Seven – Vincent Frank, Esq.

CURRENT REGULATORY ISSUES.............................67

Questions and Answers..72

Part Three – FUTURE TRENDS................................81

Introduction – Robert A. Bohrer

PERSPECTIVES ON THE FUTURE OF
BIOTECHNOLOGY..83

Chapter Eight - Dr. Clifford Grobstein

FUTURE CONCERNS FOR THE
SCIENTIST   85

Chapter Nine - Harry Casari

THE FUTURE OF THE BIOTECHNOLOGY
INDUSTRY   93

Chapter Ten - Robert A. Bohrer

THE FUTURE REGULATION OF
BIOTECHNOLOGY   101

Comment - Robert A. Bohrer

FUTURE TRENDS   115

Questions and Answers   117

Footnotes   123

About the Authors   129

Foreword

This book is for scientists who wish to know more about what business people and lawyers are doing in the area of biotechnology, for business people who want to know more about biotechnology law and science, and for lawyers interested in scientists' and business executives' views of biotechnology. The book grew out of a conference which was held in La Jolla, California on February 28, 1985. That conference, which has since become an annual event known as the San Diego Biotechnology Conference, had a rather simple, but unique purpose--to approach the emerging issues raised by advances in biotechnology from a multidisciplinary, collaborative focus.

Biotechnology begins, of course, as an applied science. However, it requires not only the efforts of scientists, but the efforts of business persons to translate the products of that applied science into what has become a multi-billion dollar commercial enterprise. As that commercial enterprise develops, it is inevitably confronted by, and interacts with, a complex political and legal structure, that offers protection to its ideas, models for its organization, and controls on the distribution of its products. Thus, the development of biotechnology is not only a scientific process, but a commercial and legal-political process as well. Yet, for the most part, despite their common interest in bringing the benefits of biotechnology to the world, while safeguarding the world from its risks, these groups rarely have the opportunity to share their perspectives, to learn of each other's concerns, and to work from a framework of common understanding. The conference was an attempt to begin such a process; to bring lawyers, business people, and scientists together, to discuss their perspectives, their views on current issues, and their views of the future.

The book is divided into three parts. In the first part, one representative of each discipline develops a general overview of his field. Dr. Ramon Pinon, a University of California at San Diego biologist and former member of the NIH Recombinant-DNA Advisory Committee (RAC), shares his recollections of the early days of genetic engineering, the issues that have been faced, and the scope of the field today. William L. Respess, the general counsel of Hybritech, Inc., one of the nation's leading biotechnology companies, discusses the many and varied roles of the lawyer in the biotechnology enterprise. Vincent Frank, a lawyer/business executive who is now president of Molecular Biosystems, Inc., deals with the broad range of issues which confront those who would bring biotechnology products from the laboratory to the marketplace.

In part two, the focus changes from the general overview to current, cutting edge issues. Dr. Theodore Friedmann, of UCSD's School of Medicine, discusses his attempts to develop a human gene therapy, research of extraordinary potential for the amelioration of a great number of genetically caused diseases. Ned Israelsen, of the law firm of Knobbe, Martens, Olson and Bear, analyzes the state of the law concerning intellectual property protection for biotechnology. Vincent Frank, donning his lawyer hat, surveys the current approach of the FDA to the approval of biotechnology products. Ray McKewon, a venture capitalist who has been involved in a number of start-ups involving university scientists, presents an interesting view of the delicate and troublesome relationship between the business world and the university.

In part three, representatives of all three disciplines turn their attention to the future of biotechnology. Harry Casari, a partner in the accounting and consulting firm of Arthur Young & Co., speculates on the future of the biotechnology business. Dr. Clifford Grobstein, UCSD professor of biology and public policy as well as a leading advisor to the NIH, looks ahead at some of the troublesome choices facing biotechnology researchers. Finally, your editor, a law professor who specializes in the regulation of risks to health and safety, attempts to forecast the future direction of the governmental regulation of biotechnology.

Footnotes have been numbered sequentially through the book; they may be found at the back of the book beginning on page 123.

>			Robert A. Bohrer
>			San Diego, California
>			August 20, 1986

Acknowledgments

Many people contributed greatly to the First Annual San Diego Biotechnology Conference and to the completion of this book. Among those deserving thanks are, of course, the conference sponsors; California Western School of Law, the University of California at San Diego, and the Section on High Technology and Computer Law of the San Diego County Bar. Among individuals deserving special thanks are Associate Dean Robert Cane of California Western School of Law, Robert Peddycord of the San Diego County Bar and Dr. Clifford Grobstein and Dr. Patrick Ledden of UCSD. I benefitted from the assistance of several students at California Western School of Law, including Geoff Carson, who assisted in the staging of the conference, Ivan Porto, who contributed to the footnoting of the manuscript, and Gary Forman, who did an extraordinary job in assisting with the initial editing of the transcripts of the conference. Mary-Ellen Norvell, ever gracious, took on the tedious job of supervising the transcription. Finally, Karen Bohrer, of the accounting and consulting firm of Peat Marwick (and my wife), provided invaluable advice and assistance at every stage.

Part One

OVERVIEW

Chapter One
RECOMBINANT DNA: CONTROVERSY AND PROMISE, A SCIENTIST'S OVERVIEW
Dr. Ramon Pinon

Recombinant DNA: (R-DNA) research, once carried out by a few scientists scattered in a few laboratories around the world, has become an essential part of a burgeoning world-wide biotechnology enterprise and a concern to national governments. This was inevitable given the complex brew of scientific, legal, political and economic issues and questions that this type of research has spawned in its short history. This chapter provides a short overview of the R-DNA revolution from its beginning to the present day.

There are two parts to this review. First, is a look back to the controversy which enveloped R-DNA research during its first decade, beginning about 1971. Some understanding of this period is useful, particularly because this experience may have conditioned our responses -- both individually and collectively, to other future issues which R-DNA research will raise. Second, is a synopsis of the current range of R-DNA research being carried out in academic and industrial settings, including some of the emerging concerns and issues which may have a profound impact on our society in the years to come.

A few definitions are in order. Recombinant DNA, as the term is commonly used, refers to the construction in the laboratory of novel combinations of DNA molecules from the same or different species. These R-DNAs can be cloned -- that is, replicated to make many identical copies, by joining (ligating) them into naturally occurring DNA molecules, called plasmids, or into certain types of viral DNAs, and propagated in an appropriate cellular host. Plasmid or viral DNA can also be artificially modified to enable them to be replicated in more than one cell type, and hence, can serve as vehicles (vectors, i.e., methods of joining) for transferring R-DNA from one cell type to another. Associated with the procedures for this _in vitro_ construction of R-DNA are a variety of methods for the introduction of such R-DNA's into microbial cells, animal cells in culture, and more recently into fertilized eggs or embryos. The combination of procedures for constructing, modifying, and propagating R-DNA's is usually referred to as genetic engineering. With these simple but revolutionary advances, genetic engineering has become a practical reality for microorganisms, it is becoming a reality for animals, and may eventually become a reality for human beings. These new developments have given molecular biologists the ability to manipulate genes, and hence life itself, to an extent and on a scale never before thought possible. This new power is, of course, also a source of profound concern about the direction in which it may take us.

THE R-DNA CONTROVERSY REVISITED

The R-DNA story effectively began in late 1971 or early 1972 when Paul Berg, of Stanford University, described procedures (which appeared complex and exotic) for cutting and rejoining DNA molecules in specific ways. He raised the possibility of techniques which would permit the introduction of DNA from a tumor virus (called simian virus (SV40)) into a laboratory strain of the human intestinal bacterium _Escherichia coli_. When this experiment had been proposed the previous summer at the Cold Spring Harbor Laboratory, immediate concern was expressed that such an experiment might be biologically hazardous and the planned experiment was postponed. The cause for concern was that tumor viruses, like SV40, although not known to cause disease in man, were capable of transforming animal cells, in culture, to a malignant state. Hence, the introduction of such molecules into _E. coli_, a common resident of the human intestinal tract and one capable of exchanging DNA with other types of bacteria (some of which are pathogenic to man), could lead to the wide dissemination of such molecules among human and animal populations with unpredictable effects.

Such concerns were brought into much sharper focus by the announcement, in 1973, by Herbert Boyer, Stanley Cohen, and colleagues, that they had devised simple procedures to produce and replicate recombinant DNA molecules from any source. The very ease with which these combinations could be made and propagated was troubling to a number of scientists. Questions were raised about the possibility that novel DNA combinations might lead to the creation of new pathogenic organisms. In retrospect, most of these concerns (which tended to be embellished with vivid scenarios of biological catastrophe) are now known to have been due to an incomplete understanding of the complex requirements needed to establish and maintain pathogenicity. Nevertheless, in the face of unprecedented uncertainties, a number of scientists suggested a moratorium on certain kinds of R-DNA experiments until a much larger group could meet to analyze the situation in some detail. This famous meeting took place in Asilomar, California in February 1975.(1) By the time of the Asilomar meeting, the few voices of concern raised initially had become a crescendo, and this new emotion transformed (some feared) what had started out as a dialogue between scientists, into a complex social issue and media "event." The significant events of the conference were the drafting of guidelines regulating R-DNA experiments, and the call for the development of safe bacteria and plasmids that could not escape the laboratory.(2) Given the uncertainties that were raised about laboratory safety at Asilomar, it was almost inevitable that the National Institutes of Health (NIH), which funded most of the work in molecular biology, would eventually be drawn into the fray. This led in 1976 to the first NIH guidelines and the prohibition of many categories of R-DNA experiments.(3)

Probably, as a few observers have commented, the mood of the times, perhaps exacerbated by the confusions and debate of the Vietnamese War and the new-found concern for the environment (the Environmental Protection Agency came into existence in 1972), created the charged mea culpa atmosphere at Asilomar. Whatever the reasons, the "spirit" of Asilomar set an important, yet paradoxical ground rule for future discussion of R-DNA experimentation: the paucity of data was to be used as a way to restrict or prohibit a particular R-DNA experiment. R-DNA experiments were a priori suspect and safety had to be demonstrated.

DeWitt Stetten, Jr., the first chairman of the NIH Recombinant DNA Advisory Committee (RAC) wrote about the Asilomar meeting several years later: "It has taken me several years to analyze and unscramble the experience of the Asilomar meeting. I now understand it more fully than I did at the time. It had many elements of a religious revival meeting. I heard several colleagues declaim against sin, and I heard others admit to having sinned, and there was a general feeling that we should all go forth and sin no more. The imagery which was presented was surely vivid, but the data was scanty."(4)

Although there was a gradual relaxation of the NIH guidelines between 1976 and 1982, they were, from their inception, a source of controversy. On one side were those who felt that the guidelines were "arbitrary and capricious," not easily defensible on any scientific ground, a bad precedent, and would inevitably result in governmental regulation of basic research. On the other side, there was varied opinion. Some felt that the guidelines were overly restrictive, but thought them the better part of political wisdom. Others felt strongly that the R-DNA experiments were dangerous. Vivid scenarios of catastrophic accidents appeared regularly in local and national publications.

In this atmosphere it was difficult to know how to proceed. What do you do when the experts disagree? Many in the scientific community who felt that the guidelines were too restrictive were troubled and frustrated by charges from colleagues and others that they were being socially irresponsible, and perhaps even a menace to society.

In 1976 the controversy threatened to leave the NIH-universities arena. Several prestigious scientists, convinced that the guidelines did not afford appropriate safeguards, began calling for a cessation or postponement of all R-DNA work. Their comments, taken up and given wide distribution by environmental groups and the media, resulted in widespread public uncertainty and unease. This eventually led the city council of Cambridge, Massachusetts, to declare a three month moratorium on R-DNA research in that city -- most of this done at MIT and Harvard.(5) The Cambridge council later set up a local committee to oversee all R-DNA work in the city.(6) Other communities began to think seriously about following Cambridge's lead. San Diego, as I understand, was one of several cities that set up a subcommittee on Recombinant DNA. Concerned over these developments, even many scientists began to think that a federal bill regulating R-DNA work would be preferable to a hodgepodge of local regulations.

In 1979, I became a limited participant in the R-DNA controversy when I was asked to serve on a much-expanded Recombinant DNA Advisory Committee (RAC). Pressure to expand the RAC to include non-scientists had increased and finally became impossible to ignore. The first meeting revealed many of the weaknesses inherent in a committee whose members come from very different backgrounds. Deliberations were difficult at times, not only because of large differences in scientific expertise, but more importantly, because of the different perceptions that the members brought to the committee. Questions of whether the members represented different constituencies were continually raised. The differences between those who saw the guidelines as an interim measure (mainly scientists) and those who felt that they should not be relaxed were never resolved, and an uneasy tension prevailed at most meetings.

In spite of this, the Committee continued to tackle the basic assumptions upon which the guidelines were grounded -- i.e., the conjectural hazards of R-DNA work. Gradually, a progressively changing assessment by the scientific community of these conjectural hazards led to a relaxation of many of the more stringent guidelines. As Donald Fredrickson, director of the NIH, put it: "No new facts or unconsidered older ones have emerged to support the fears of harmful effects, and one prominent early proponent (J. D. Watson) of the guidelines has repudiated his support for them. At the least, there is growing sentiment that the burden of proof is shifting towards those who would restrict R-DNA research."(7) This shifting continued to accelerate, so that by 1982, more than 90 percent of all R-DNA experiments were exempted from the guidelines. A significant exception was the provision which dealt with the deliberate release of R-DNA molecules or organisms into the environment. And it is this provision which forms the basis of a recent suit to halt dissemination of genetically engineered bacteria into a potato field at UC Davis.(8)

The R-DNA controversy, which had raged for a decade, ended in 1982 with a whimper. To be sure, a few uncertainties remained; but except for a few fringe groups, no one any longer seriously believed the catastrophic scenarios previously envisioned. W. J. Gartland, director of the Office of Recombinant DNA Activities, summarized the feelings of many:

> It is possible that the recombinant DNA affair will someday be regarded as a social aberration, with the guidelines preserved under glass. Even so, we can say the beginnings were honorable. Faced with real questions of theoretical risks, the scientists paused and then decided to proceed with caution. That decision gave rise to dangerous overreaction and exploitation, which gravely obstructed the subsequent course . . . the lessons learned here should help us through the turbulence that is sure to come.

The gradual relaxation of the guidelines shifted the focus of discussion of R-DNA research from concern over the hazards to the potential benefits of harnessing genetically engineered organisms, viz., the making of products of commerical or medical importance. These possibilities, given great impetus by a decision of the U.S. Supreme Court, in June 1980,(9) that forms of life carrying a manmade genetically engineered component can be patented, have spawned a multimillion dollar industry. However, commercialization of genetic engineering also has not been without its critics. Much of the criticism is directed at scientist-entrepreneurs who, recruited by offers of stock options into new, small-venture capital companies, are accused of behaving unethically -- exploiting public resources for private gain, or announcing new results at press conferences to maximize their companies' stock values. For some, the shift in concern from biohazards to the value of R-DNA stocks is a measure of how low we have sunk. For others, it is simply an indication that molecular biology is growing up, and like its sister branches of science -- chemistry and physics -- is evolving an acceptable association with industry.

R-DNA WORK IN THE 80s

I think it is possible to divide current R-DNA research into three categories. Although not mutually exclusive, these are nevertheless sufficiently different in their objectives to merit separate discussion.

The first is that which traditionally has been seen or thought of as basic research. This type of work is carried out almost exclusively in an academic setting, and its objective is to analyze a bewildering variety of cellular processes at the molecular level. This encompasses an enormous range of studies ranging from those on the simplest viruses to those on the most complex cellular interactions. This work is unpredictable, open-ended, and tends to progress discontinuously. Particularly important areas of research in the last two or three years are those seeking to understand how DNA is replicated, and how the expression of genes is regulated. These areas of study have literally exploded. We have learned about and isolated the DNA sequences which are necessary for replication; we have characterized sequences referred to as "promoters" and "enhancers" which regulate gene expression; we have learned how antibodies are synthesized, and are beginning to understand how the cells in the immune system work and interact with each other to help vertebrates protect themselves from a wide variety of chemical and biological agents; we have made tremendous progress in understanding how viruses cause disease, and are beginning to identify the genes which may be responsible for cancer. These studies would have been inconceivable without the use of R-DNA techniques. Ironically, many of the studies, particularly those relating to animal viruses and cancer were possible only after the relaxation of the NIH guidelines.(10)

The second category of R-DNA work takes place primarily in the industrial setting -- in the laboratories of the large, transnational pharmaceutical companies and increasingly in a large number of small biotechnology companies, most of which are backed by large corporations. This is applied technology -- the exploitation of R-DNA technology to produce products or organisms that have commercial or medical significance. As with basic research, a wide range of organisms and approaches are used. But unlike basic research, the work tends to be mission-oriented and uses methods or techniques already developed. In many instances the research is carried out as a collaboration between industry and academia -- a relationship which is not without its problems. A number of questions about the role of the scientist-entrepreneur, trade secrets, proprietary information versus freedom to publish, rights and terms of ownership of ideas, or organisms, or methods -- are now being asked.

This aspect of R-DNA work shows tremendous promise, but has had few concrete successes so far. We expect some areas may payoff sooner than others. The scope of these projects may be appreciated by a few examples: construction of microorganisms that may be useful in cleaning up various types of environmental spills -- ranging from oil products to toxic chemicals; construction of bacteria to produce a wider variety of antibiotics more economically; construction of a variety of cell types that will produce hormones and proteins that either cannot be produced any other way or cannot be produced easily and economically. (Examples of this last are insulin for the treatment of diabetes; interferon for viral infections; factor VIII for hemophiliacs; a 1 - antitrypsin for emphysema; renin for the production of cheese; vaccines for the treatment and prevention of viral and parasitic diseases which affect people and animals; improvement of plants of agricultural importance to increase yield, nutritional value, and stress tolerance.)

The third aspect of R-DNA research, still in its infancy, is the application of genetic engineering to human beings. Although at present these goals are stated in terms of interventions which will correct genetic defects, interventions aimed at enhancing "normal" people can be and are being imagined. We are clearly in unchartered territory here since we have great difficulty in defining what "normal" is, and determining what type of enhancement is desirable or acceptable. These latter possibilities have raised questions substantially different from those that were raised in the early years of R-DNA work.

No one envisions global catastrophes, but the concerns reveal deep-seated anxieties that work in this field might reconstruct human beings, or that the human life may be devalued. Our deeply held feelings about what a human being is seem threatened. These are complex and difficult issues. However, the lesson from the early R-DNA controversy -- that there is no substitute for careful and informed discussion -- should not go unheeded.

Chapter Two
A GENERAL INTRODUCTION
TO THE NEW BIOTECHNOLOGY BUSINESS
Vincent Frank, Esq.

I've been asked why I became involved, as a businessman, in biotechnology. Aside from the simple answer of profit and increasing the value of one's own net worth, I think there's an emotional factor: this is groundbreaking material. It's new technology; it's futuristic and it holds the promise of good things for mankind. So, as opposed to investing in something like fast food restaurants, investing in biotechnology calls for a commitment; a vision of how the various biotechnologies can improve mankind's lot in the future. Vision is an important prerequisite.

I've also found that one must have the sensitivity of a pachyderm, because there will always be detractors. Someone will always say, "This technology hasn't worked yet. The products haven't been produced yet." And in point of fact, five years ago if you listened to people talk about what biotechnology was going to do in the next five years, we would now all be living 50 years longer and our energy would be produced from little "bugs" that turn our garbage into kilowatts. These and similar predictions have not yet happened. However, this does not mean that the future of biotechnology as a business will not be a great one. So, let me share with you some observations I have of why one gets involved, the thought process that one goes through, and point out some issues that may be of interest regarding this new and promising field.

There are many reasons for not getting involved in biotechnology. Like any unchartered territory, the fear of the unknown keeps people away. Also, strategically, there is something to be said for letting someone else go up the learning curve. Now when I talk about biotechnologies, I'm not just talking about recomminant DNA, i.e., cloning. There is a wide array of biotechnologies. There's cloning, monoclonal anti-bodies, DNA probes. There will always be those who jump on the bandwagon after someone else takes the chances and does the tough work. But fortunately, there will also always be pioneers to get involved at the basic stage. I will talk about that type of biotechnology company which starts a technology and works its way through to product development.

The first problem is identifying the technology: What is it? How well developed is it? I distinguish technology from science. The discovery of the double helix was science. The ability to manipulate that double helix, to do various things with it, that's technology. And the next step, (engineering, if you will) is taking that technology and embodying it in a product that is useful and widely available. This is the main business consideration when looking at a technology. What can it do? What needs to be done to develop that technology into a product that is going to be useful to people, e.g., a diagnostic product, a therapeutic product, something that increases the productivity of your food supply? How much is it going to cost and will the results be readily commercializable? Are your expectations realistic? One mistake that is often made is underestimating the amount of time and money necessary to bring your technology to market. If someone says it's going to take a million dollars to develop the breadboard of a device or the prototype of a diagnostic kit or therapeutic -- you can almost bet it's going to take two to three times that much (and probably take two to three times longer than originally planned).

I'll illustrate all this by using a personal experience. My company is involved in developing DNA probes for in vitro diagnostic purposes. A probe is a small piece of DNA that is designed to hybnilize to, and thus detect, specific complementary DNA molecules. Now, the technological ability to create and modify these probes holds great promise for medical treatment. If you can detect the DNA of various disease sources, e.g., viruses and bacteria, you can determine if the patient has a particular disease with great specificity. But how do you develop this into a product which somebody in a clinical laboratory or physician's office can use in fifteen or twenty minutes? Something that's inexpensive, easy to handle, and that is preferable to current methods of diagnosis. This is not a trivial question. You find that scientists are visionaries and they will tell you what science or technology can do as if it's already doing it -- as if that quantum leap has already been made. Such, however, is not usually the case. There's a host of problems in developing that technology. And once the technology and its potential are identified, then more practical considerations come into sharper focus.

One of the most overlooked aspects, for example, is, who owns the technology? Where was it developed? Who developed it? There aren't too many Leonardo Da Vinci's out there who sit in their studies at home and invent things. Many inventions come universities that have programs sponsored by the government or by private industry. Often in the excitement to get a venture going, entrepreneurs get deeply involved with the research and development only later to find out that there are encumbrances on it. This is especially true where the entreprenuer is a scientist who usually is not thinking whether Hoffman LaRoche or Abbott Labs funded this project, or whether there was a grant involved. He or she is interested in the pursuit of knowledge. But ignoring the source of the technology and of its development can cause problems later on in attracting subsequent commercial investment, let alone potential lawsuits for patent infringement.

So university or government connections may be sources of major problems concerning biotechnology development. These connections are related. A common scenario is that the founder or employee of a biotech enterprise is a scientist who developed a particular technology and now owns significant equity or stock options in the company. The company is ready to go, but it suddenly finds out that this particular technology was developed under an NIH grant, in the founder's laboratory at the university, along with someone else who has since left the university and is at another institution. So you have to unravel. All these complications of ownership must be sorted out before one invests hundreds of thousands or millions of dollars in the further development of this technology. This is a very significant factor in determining the valuation of your technology. For example, if a particular invention was made under a grant to a university, the government usually places limitations on the exclusivity of that particular invention. If a company wants to develop or subsequently license it, it will have to pay royalties, and may still have to worry that its license is going to be non-exclusive. So the relationship of technology development at a university and its transfer into the private industrial sector is not often a desirable one. Financial people who have already done their homework in assessing a nascent biotech venture, in fact know that the technology is likely to be that company's main asset.

It is also frequently the case that a large company is a sponsor or benefactor of the research giving rise to the technology. Sometimes this is even worse, because in typical fashion, the large company will fund the activities of a particular university laboratory in return for a right of first refusal to all inventions coming from that laboratory. If another company is interested in investing hundreds of thousands of dollars in acquiring that technology, it can signal the sponsoring company to claim its ownership interest. This can actually depress further development of the technology if the sponsoring company chooses neither to develop the technology or relinquish rights.

In any event, once the issues of the technology identification and ownership are resolved, development itself may proceed. Initially, we look to the technology's commercial applications. By this I do not mean an application which is useful and interesting, but rather one, albeit less exciting, which has the potential for great profit. This is a biotech company's, and for that matter all for-profit companies, ultimate raison d'etre. Although this may appear crass to purists, this is the only way to attract the huge investment of the dollars necessary to develop a product or process for market.

One of the major problems confronting health care R & D is the relatively long time period involved. Again, five years ago people were saying that in five years there were going to be $5-10 billion in sales from biotechnology products. That just hasn't happened yet. The whole scope of biotechnology business, (if you took all the money invested in it and all of the sales from it in the last 5 to 7 years) would pale in comparison to an average sized chemical or drug company, with a couple of billion dollars of sales. It currently is not a mature business by any means. Many biotech companies virtually exist on research contracts and interest from their inverted cash. Few even operate at a profit. Now, the fact that this is the case doesn't necessarily mean that biotechnology is a bad business. It just means it is a very young business which must be pursued realistically and patiently.

I always hear biotechnology analogized to the computer industry. They say, "Oh, yeah, that's like computers." True, it is a new technology, and one which promises broad applications. But the computer industry does not have some of the impediments typically encountered in biotechnology. For example, in biotechnology one must deal with some issues which are highly emotional. The very mention of genetics or manipulation of DNA particles of life, get certain people agitated.(11) It's a very emotional issue -- particularly because there's so little public awareness about it. The regulatory bodies at all levels are very cautious about their examination of biotechnology products and their clearance for testing or marketing them. So, the development time is much longer.

Also, you're dealing with an ever shifting technological foundation. Almost on a daily basis, new things are emerging which make the old technologies obsolete, or that make something once only possible in a week's time, now possible in one afternoon. Take R-DNA, for example. The 1981 headlines read: GENE MACHINE SYNTHESIZES DNA OVERNIGHT. Now you can do it in an afternoon, and in quantities that can last years. So the technology is rapidly moving forward and your company always has to keep one step ahead.

Resource allocation in a company is also a critical area. A young biotech enterprise often bursts out of the gates and creates the huge superstructure of a mature company. I think caution here is the better approach. Develop what you need at that particular stage of your existence. Don't go out and hire 200 people, only to find six months later that your capital is dwindling and your product development isn't on schedule. You're now going to have to fire all those people, sublease space in your building and/or raise money at a less than opportune time. Tight control and allocation of resources is extremely important.

Project control is needed. Scientists and technical people must think about the commercial goal. Though this attitude may conflict with the scientific investigator's perspective, one must understand that if you develop something, e.g., an inexpensive cure or vaccine or diagnostic, it will provide the greatest benefit to the greatest number of people only after it is a marketable product.

In determining commercial potential, many of my colleagues in the industry typically ask, "How do you figure out what the future market is going to be for something for which there's no current market?" This is a very difficult process. In trying to develop a superior substitute, analyzing the market is a rather straightforward process. You can easily determine what the market is for a faster, more accurate type of hepatitis test, for example, if you examine the current products market. But suppose your company is developing something which will detect a genetic abnormality and for which there is currently no test. What is the market for that? Even if you can identify a number of current cases, how many more would use the test if it were less expensive or more accurate? This is one of the most difficult assessments a biotechnology company makes. Consequently, a lot of biotechnology companies concentrate on those things that they can most easily quantify. And, in some ways, that is a good approach, because if you develop something successful that is market-tied, or market-driven, then you can have a bit more slack with which to work on things which aren't market-driven. But, you do run the risk of being out there with a "me too" product, and possibly shortchanging your future. It's a difficult dilemma.

Another factor to consider, and one that is often given short shrift, is competition: Who is out there doing what? Biotechnology has become a very secretive business. University people who used to engage in free exchange of information no longer talk to their colleagues in industry because they're afraid they will take that idea and use it for their own commercial benefit.(12) Biotech companies are paranoid about exchanging information. You cannot have lunch with someone unless you have a "confidentiality agreement"! So, it's really hard to find out what the competition is doing. Competitors usually don't have products on the market that you can inspect and dissect. But using whatever sources you can (there are various services and marketing publications which do cover various biotechnologies, and which are getting very specific at keeping track of what the competition is doing) is very important. It helps you steer development and maintain that sense of urgency, to know what the other guy is doing. Not just in your own technology, but in other areas that may give rise to products which do what your technology does.

    I would like to make a few comments about the operation of a biotech company, especially a young company, and some of the personal interactions which may be important. A company should define itself. In the early 1980s, when there was a great flood of new biotech companies and initial public offerings, this was the time Genentech went public and there was a feeling that, as a company, you had to get involved in everything because you had this big pot of capital and this promising technology. Companies now realize the need for greater definition, the need to focus. It's an extremely important point. From my own personal observations, I have noticed a tendency, early in a company's development, to avoid putting all your eggs into one basket. That may be a good approach if all your eggs are fertile, but doing so may spread resources thin and drag down the enterprise. Define and focus. A flexible definition and occasional reassessment of where you are going (and whether you should go there) is helpful in coping with a technological foundation that often shifts. It is equally important to get people to rally around a defined goal. I like to look at it this way. If someone in your company is asked at a cocktail party, "What company do you work for and what does it do?", I want them to be able to answer in one short sentence. I think this is extremely important, because biotechnology is a very people-oriented business. The people working in biotechnology are highly trained and are very motivated when they know the ultimate goals. People like to feel they are vitally involved and having definition and focus helps.

Finally, I'd like to say a few words about financing. It is no secret that when you're involved in the development stage you need ongoing capital -- ongoing financing. One question that often arises is, "What is my company worth?" The simple answer (and I guess it betrays a certain cynicism I have about the investment community) is whatever somebody is willing to pay for it at that point in time. In 1980, if you were a biotechnology company, people were pounding on the door. People were calling their brokers and telling them to buy as many shares of biotech as possible. In 1982, you couldn't get anyone to return your phone call if you were associated with a biotech company. But in 1983, there was renewed fervor. In 1984, we were in another quiescent period for biotech initial public offerings and R&D limited partnerships. But 1985 and most of 1986, biotech valuations sky-rocketed and public offerings abounded. The momentum then halted abruptly in July 1986. Do you discern a pattern?

There is a necessary attitude concerning biotechnology investment. Don't fall into the trap of looking at how many employees a company has, or how many patents, or how many products. Real valuation is based upon product sales and profits which are consistent over a number of quarters. On retained earnings. On the bottom line. Anything else only addresses how the industry in general, or your particular technology, is perceived by the investment community at a given point in time. So, until your company is making consistent product sales and putting money into retained earnings, valuation will be volatile, and you better hope that the biotech investment attitude is cyclical enough so that the next up-phase will come before you need more money!

Chapter Three
BIOTECHNOLOGY: THE LAWYER'S ROLE
William L. Respess, Esq.

Dr. Pinon and Vince Frank have presented an overview of the kinds of scientific and business considerations which are characteristic of the industry we refer to as biotechnology. And, this means to me more than just recombinant DNA. It is also genetic probes, monoclonal antibodies, and a range of other new technologies as well. And, I'm not so sure that it's fair to leave out the old-line pharmaceutical companies. They've been using organisms, for example, to help them manufacture their products for many years. And, I guess, the brewmeister thinks he's in biotech also, because he makes beer with organisms too.

In any event, since I became involved in the biotechnology industry as a lawyer about six years ago, when Genentech came to Lyon & Lyon as a client, I've felt that I'm sometimes the unwelcome guest at a party. You have to invite your lawyer, but you hope he'll come late and leave early. Much of that adverse view of lawyers is reflected in the lay press. Recently, I read an article in the Los Angeles Times, actually from the Op-Ed opinion, a page written by a lawyer, and entitled, "Should Lawyers Prosper From the Technology Under Stress?" The lawyer was commenting on the nuclear power industry and regulation. He maintained that Congress and the judiciary were the primary problems with that industry. But, he didn't leave the lawyers out. In fact, he made the following comments: "I suggest there exists a third unindicted coconspirator in the murder (I like to call it delusion theory) of nuclear technology, and that is the practicing bar, of which I am admittedly a member. From everything that a non-technically trained person like myself can discern from the literature, the western Europeans, the Japanese, and our neighbors to the north in Canada all seem to be making a go of nuclear fission. We in America did so for a number of years, but now apparently we cannot. What's the difference? I suggest the difference lies in the ratio of lawyers to technicians, scientists and engineers. In Japan, there are no more than 15,000 lawyers;(13) in the United States, 35,000 law degrees are awarded every year. In the nuclear power field, the large numbers of lawyers representing all sides have lawyered it to death. We have reached the point where, today, assuming an American utility executive was foolish enough to build a nuclear power plant, it would behoove him or her to make sure the company had a first-class legal team before it began to acquire good engineers."(14)

The biotechnology industry is also an industry under stress, but I don't think this is due to possible imminent failure, such as the nuclear power industry may be facing, because it has outlived its usefulness in some people's minds, or perhaps because of overregulation. I think, to the contrary, it is under stress because it is suffering from a high level of success. As companies, we are competing and will compete for the forseeable future, for a limited reservoir of talent, a limited pool of investment capital to fund our enterprises, (Vince Frank touched on that), and we must make decisions whether to commit to products in a future market which we cannot know. In some cases you may not even know if somebody else won't get there with a patent position which might freeze you out. And so, companies get over-stressed in terms of planning their activities.

Biotechnology does differ from nuclear technology, but there are parallel aspects. For example, nuclear technology makes weapons of war. I suppose that ultimately somebody will turn biotechnology in that direction as well. We too are concerned about environmental hazards, and all kinds of regulatory pressures. The adversary system has resulted in substantial lawsuits already. We have patent infringement suits, activities to enjoin agricultural experiments with engineered organisms, and we are concerned about proprietary interests, e.g., technological advances which may involve tissue taken from a patient. Who owns the benefits -- is it the patient or is it the researcher himself? We have employment litigation suits by former employers seeking to enjoin the new employee from hiring a member of its research staff. Unfortunately, Hybritech, a leader in biotechnology advancement, has also been a leader in patent litigation. We have been the plaintiff in one suit and the defendant in another. We are the plaintiff in an antitrust action.(15) We have been the defendant in employment litigation.(16) So the adversary system is a significant part of our everyday life.

I have the feeling that our pioneering in this area will, however, be followed by other companies, perhaps Molecular Biosystems one of these days, or Synbiotics and other companies here in San Diego. Cetus is already there, as is Genentech. J&J (Johnson & Johnson) has sued Becton Dickinson,(17) and there may be other cases filed yesterday that I am presently aware of.

I think that's the bad side of the lawyer (and hopefully all of it) in biotechnology. I think the real role, however, that lawyers will play, the most important role, is a positive one, which will go unsung, because it will not be in the public eye. What will be much more important will be lawyers working behind the scene to prevent problems, anticipate them, and also, I think, sometimes to make the difference between the biotechnology enterprise surviving on the one hand, or actually prospering, on the other.

When should you get your lawyer involved in your biotechnology enterprise? In my opinion, he or she should become involved before you have the enterprise. In that respect, most of the enterprises that are represented by business people here today, have just been formed by a combination of academics and business types, with ideas on the one hand and perhaps capital or the ability to raise capital on the other. They do not have a corporation, they do not necessarily have a business plan -- they'll go to one expert or another to help work those things out. But before they ever incorporate, I think they need the services of a lawyer. And this is for more than just preparing and filing the incorporation papers. The new venture needs to be counseled in how the original partners are going to work together. Are they going to be employees? Are they going to be board members? Who is going to be the president? Will they be all of these and Chief operating officers all rolled up into one? Who is going to be the lawyer, etc.? Those kind of things are important to work out at the start, because who knows who is going to want to be president after the corporation is already formed, stock is issued, and people have started buying that stock. Lawyers have, over the years, acquired a lot of experience and insight into enterprise organization and launching. You should have a talk with your lawyer before you ever get out of the starting blocks.

Vince Frank alluded to the fact that an issue of importance to the venture capitalist or the source of capital is ownership of the technology that you are going to exploit. I think the situation is often so complex that it would be foolhardy not to get a lawyer to act on it early. When you get into the situation of raising money publicly, your investment bankers are going to insist on it as an element of due diligence anyway. It would be terribly embarrassing to launch a new venture, spend a lot of money, and find out the government has granted a non-exclusive license to Abbot Laboratories which has a billion dollars to spend on the project compared to your 500 thousand start-up fund.

After the new venture is started, everyone is working together, at least in the early days, with a view toward making that first breakthrough product that will demonstrate a real technology and not something that is merely a gleam in someone's eye. The lawyer's role begins to expand geometrically. The facts came home to me the other day when I saw Genentech's price going up very rapidly on the stock market. It is now valued at over a half-billion dollars by the market valuation test, and probably has seven or eight hundred employees. But it is not profitable in the sense that most businesses are looked at. It doesn't make a profit on product sales, but does so from research contracts and from interest earned on its CDs, as Vince Frank was pointing out.(18) So how can a company that big get along without some sort of legal capability? In fact, I think it probably cannot. In the early going, the most modest venture has to have at least some legal facilities.

Some of the major legal issues are really self-evident, but let me enumerate them anyway. You have to comply with local regulations and state and federal regulations concerning handling of toxic materials and living organisms and other research ingredients (radioactive materials can be shipped in commerce but only under certain circumstances). The company is expected to comply with those regulations, and most new venture people, at least until they get in the hands of more sophisticated business people, have no idea of the burdens they are going to face once they start that business.

Early on, you have to make a decision about protecting the technological base that you started with and that you hope to acquire. You need some insights as to the relative advantages and disadvantages of trade secrets on the one hand and patents on the other. How do you build a security system that would convince an outside observer, (in any later litigation over trade secrets) that you had actually taken reasonable steps to protect your information. Lawyers, because they have, for the last fifty years, litigated trade secrets, have a great deal of insight as to how this ought to be done. I will leave the patent matters to Ned Israelson this afternoon, but I think you'll agree that that's probably been the sexiest topic in biotechnology in the past few years, beginning with the Chakrabarty(19) case which Dr. Pinon alluded to.

What about employment agreements? It is amazing the pitfalls that exist with respect to employment of technical people. California, and a number of other states, have begun to limit the rights that an employer can acquire from an employee, by statute. Those rights are probably as broad as an employer can have; but the bad side of it is that without a properly drawn employment agreement, the technology acquisition section of it may be held unenforceable, and you may be unable to compel the employee to turn over anything. So, it's not a good idea to xerox the old Baxter-Travenol contract, put your company name on it, and go ahead and launch your new enterprise.

When you get into the hiring of new personnel after the formation of the enterprise, people who didn't participate in the forming of the enterprise itself, you are going to have to deal with a lot of serious problems which require some kind of legal oversight. You are probably going to recruit those people from successful enterprises, bigger companies, where they have already made a track record. Improper recruiting techniques result in lawsuits - there is a proliferation of these lawsuits in the

Silicon Valley. They have begun to find their counterparts in biotechnology. I alluded to two suits in which Hybritech has been a participant -- unfortunately on the sued side and not on the side of delivering the complaint. It is expensive litigation and it can hurt a small enterprise. There are safeguards which can be built into recruiting policies, which can minimize the risk to the startup company. It would be foolhardy, I think, not to have the foresight of doing something about that. The significance of this area for potential problems is illustrated by a seminar I attended a few weeks ago. It was a two-day seminar for lawyers, at which this point was the sole topic of conversation. Lawyers were gearing up, and there were both plaintiffs' and defendants' lawyers there, I suppose they each know the tricks of the trade the other is going to employ.

When you hire your key man, he is going to want golden parachute protection, which may involve some exotic employment agreement: stock options, even guaranteed employment, if things go sour after a few years, until this person can get himself back on his feet in another enterprise. If the company goes public, you have to comply with many SEC regulations.[20] Before you go public, you want to be aware of the extent to which you will have to undress in public by disclosing the perks that your key employees have. Sometimes your other employees do not know, and they will want to know why they do not have the same kind of advantages. So think about those kind of things.

Again, in California, the employment relationship is now burdened by a host of new legal doctrines. California seems to lead the way in almost every new fad, including now, employee wrongful termination lawsuits. As a practical matter, you'll need a lawyer advising you as to your employment policies, and termination policies, to avoid litigation. There is now an implied covenant of good faith in dealing with your employee. You cannot fire at will anymore, only for cause. If you have a generous supervisor who gives someone the highest rating until he or she finally gets to the point where he or she can't stand someone anymore and fires this person, you probably don't have a basis for termination in California. Because on the record you had an outstanding employee there.

The policy manual you brought from Baxter-Travenol (which was photocopied and on which you substituted the name of your company) may now have so many complex policies in it which you will have to follow (and which you didn't know were there), that without realizing it, you may deprive your employee of corporate due process, and find yourself the victim of a lawsuit.

Often, you'll find that just after the company goes public, everybody wants to make a killing on that stock that has been held for so long. But then you are into another area of SEC regulations, which involve insider trading.(21) Any start-up company that doesn't have a lawyer overseeing the purchase and sale of shares by key employees is asking for a stockholder lawsuit. And it is not just officers and directors who must be supervised, but also that key researcher who knows that a big breakthrough will be announced by the company tomorrow, one which heretofore has been held secret, who has his broker buy a hundred shares or five hundred shares or whatever. He is still an insider, and the implications are just as great for him as they would be for anyone else. So you don't just call your broker. You better first call your lawyer. These things, can be managed in a way which avoids those kind of problems.

A new venture's initial capitalization will probably run out in a year at most. As Vince Frank was saying, the money goes a lot faster than you think and makes for a lot less progress than you had hoped. So it may be too early to go out to the public to try to raise money. You are going to have to go to a partner, either a venture capitalist, another corporation, or a bigger corporation with lots of cash, in order to sustain the company until it has the kind of success which will make it possible to sell stock to the public. You have to have an agreement with these companies; either venture capitalists or a corporation that is in a related technical area. This is the next area where fledgling companies probably face the greatest risk to their future prosperity. Because if Big Brother smothers you, you're not going to be a big, successful company, although you may survive as some sort of legal entity. There will be pressure along the lines of: "In return for our money we want equity participation." Questions have to be resolved as to whether or not that equity can be diluted, therefore making it possible to raise additional money from other people who do not wish to be a co-investor with a large corporation. It may take the form of an exchange of technology that is so restrictive that it will not leave anything for the new company to do. Remember, the new company has a narrow base. If it gives up all of the rights to the improvements of that narrow base, without at least an option to expand and acquire some of those improvements, it may effectively restrain its ability to mature as it would like otherwise to do. It will have lost the flexibility to go forward. In that kind of situation, it is my feeling that the new company (and perhaps I'm naive in this regard) should, in effect, write out its game plan and then find an investor. It is easier persuade an investor into accepting some modification of your version, rather than trying to talk him out of his own preconceived ideas of how the deal should be structured. With your lawyer at your side, you can anticipate some of the problems that you might face and dispose of them initially.

In these agreements, you have to think about long-term implications. If you are going to develop a product and manufacture it, and you are given free rein, your partner may expect you to pay the freight in patent litigation and product liability lawsuits. Is that really fair to the small company? It is something which must be negotiated. If it is not provided for, it may result in an implied obligation which should have been dealt with expressly.

I mentioned rights to program improvements. You may want your agreement reviewed for compliance with the antitrust laws. In fact, The U.S. Attorney may be interested in some of your agreements. The new companies have a host of regulatory burdens, which will be addressed in other aspects of the seminar today. In addition to the FDA, there are the EPA and OSHA requirements. I've mentioned the SEC. All of these regulations are intended to serve some public good. The paperwork they now generate is just enormous - it is incredible what the bureaucrats have been able to crank out in the public interest.(22)

I think a good example of the kind of regulatory scheme that was conceived in good faith, is the one that relates to OSHA's new requirements that hazards in the workplace be identified to the employee so that he can avoid those hazards or seek proper medical care.(23) By the end of this year, you must label all your products that you sell with some other manufacturing entity, with all the hazards that they contain. By 1986, everything in your workplace has to be labeled with all the hazards they display or present. You must educate your employees, and be prepared to respond immediately to a health care professional. You cannot claim trade secret protection if she asks you what was in that flask that spilled on her patient's hand. (But you can get an agreement from her retroactively to keep the information confidential.) The industry group that opposed the breadth of the regulations indicated that it would cost $4,000 an employee to comply with these regulations. I don't know if lawyers can keep the price down -- they'll probably add to it -- but, you need one just to get through this minefield of regulations.

One surprising area of regulation for the biotech lawyer is the extent of regulation by the federal government of exports from the United States.(24) In fact, all proprietary commodities and technology in any exportable form, including trade secrets, have to be licensed by the U.S. government before exported from the United States. Fortunately the government recognizes a class of license known as general licensing, which can eliminate the need for a specific application. But, if you attempt to make a deal with a foreign company, you ought to be ready for these regulations. Now, how can you export technology? A visitor from a foreign company, including your own foreign subsidiary, comes to San Diego, goes to your plant, and leaves the country. Result? You've exported technology. Because, whatever he saw, or whatever he learned in discussions, to the extent that it is proprietary, is covered by those regulations. This may not be a problem (unless you intend to deal with certain eastern countries) if you have acquired written assurances that he will not re-export the technology. This means more than that he will not defect to the Soviet Union. It means he can not engage in a business deal with another country.

Now, what are some of the other prohibitions against export? In addition to confidential information, there are those things in short supply, and considerations of national security or foreign policy. We cannot trade at all with North Korea, Cuba, Vietnam and Cambodia, and to a certain extent, Libya. (Fortunately, Libya is entitled to receive medical products, however, for humanitarian reasons.) Products of biotechnology which have military or police application cannot be exported to the Soviet Union and a host of other eastern countries. Police applications cannot be sent to South Africa because we do not approve of the apartheid policies of that government. So, if you otherwise think you have a legitimate transaction that can be carried out in South Africa, you better be sure that you do not run afoul of the export regulations, if that technology can be diverted to police application.

How did I learn all this? Hybritech exported an instrument with an onboard microprocessor, the kind that's in a handheld calculator. They were seized in San Diego by customs officials, under Operation Exodus, which is designed to keep helicopters from North Korea. They wouldn't give them back. We had to go through the regulatory process and get the license to get them back. In other words, they would not release them to us even for domestic use. We were caught -- there's no flexibility in these regulations. They are basically doomsday-types. For a small company, whose total inventory of foreign products is sitting in the airport at San Diego for six weeks or so before the bureaucracy can be moved for its release, this can be devastating.

The computer industry has been the big target for this kind of treatment, because it is that kind of miniaturization which is most useful for military applications and about which the Defense Department has been most concerned. Now at least I am informed. Having been burned once, we looked into the extent to which they might regulate the transfer of hybridomas and so forth, and we got the bureaucratic stall. In other words, no answer -- they wouldn't say that they were covered or that they weren't covered. The Department of the Defense is looking into products of recombinant DNA. I would not like to be the lawyer of the company whose president is arrested on returning from Europe because he had a briefcase full of confidential information, and no written assurance that the company he talked to in Switzerland wouldn't reexport it to the Soviet Union. He may have to spend a few unexpected days in New York, clearing Customs.

Let me just touch on some of the other areas where lawyers play a role at Hybritech and at other companies I know. We have to worry about trademarks. In addition, if you produce product literature or ad copy or press releases, you have to be careful not to state something about your product that a products liability lawyer could use against you in some way; or a lawyer for a class action lawsuit on behalf of the stockholders could use against you. Businessmen are basically optimistic and lawyers are equally pessimistic. For that reason, they tend to provide a useful counterbalance. If you are going to make a public offering of stock or a research partnership (and I think probably the latter has had its day as an effective money-raising vehicle), it is effectively impossible unless you have an unholy alliance of your lawyers, investment bankers and accountants to pull it off. You might as well sit back and enjoy the ride, because they'll do all the work. You won't be able to do anything, really, except write the business plan for them.

If you have a successful company that is publicly held, you may want to think about shark repellent for unfriendly takeovers. Shark repellent is the current name for putting in all kinds of protective devices in your articles of incorporation or by-laws, to make it difficult to take over your company.

The list goes on. My point has not been to present a "how to do it," because I think that's a topic for another day or days. I have tried to raise your awareness of the kinds of problems that can be encountered. I think I'd rather see my lawyer in my office, rather than in the courtroom. For that reason, I think the sooner you get one on board, either as private counsel or as in-house counsel as is more likely, the more likely it is that you will to avoid significant problems.

COMMENT: LAWYERS, BUSINESS PERSONS AND SCIENTISTS:
BIOTECH PROBLEM SOLVERS
Robert A. Bohrer

In reviewing the remarks of Ramon Pinon, William Respess, and Vincent Frank, I began to understand the relationship between our basic perspectives. We are all problem-solvers at base. Scientists deal with knowledge problems, business people deal with the problems of raising capital, making marketing decisions, making management decisions, and lawyers deal with a variety of complex people and social institution problems of the type we just heard about. We are all problem-solvers and all of our problems are, in fact, involved in technology development. That is, technology development begins with knowledge problems, but it doesn't end there. If it ended there, it would end in the laboratory. For technology to grow and become, as Vince Frank said, "engineered and commercial," it requires the joint efforts of knowledge problem-solvers, of management problem-solvers, and social institution problem-solvers.

QUESTIONS AND ANSWERS

Question: I have a question for Vince Frank. When and how did Molecular Biosystems evolve and when and how did you get involved with the company?

Answer: Frank: I got involved with MBI as a lawyer -- as their private counsel, e.g., I did their articles of incorporation. When we started out, it was basically three people in Chicago talking over cocktails about starting a company. There were two physicians who had invented something. They wanted to license it, and wanted to know if they could or should start a company. My background was corporate finance and securities. After we found someone who was interested in financing what was at that time a concept, and got a commitment for the initial seed capital (which was in the form of a private stock placement), the first thing which I admonished my clients to do, who were by that time the two doctors and two Ph.D.s, was to hire a couple of business people. They and the investment bankers involved talked me into serving as the Vice President -- Finance, Law and Administration. I took a leave of absence from law practice to get involved in the company, and have since lost the opportunity to return to my private practice in Chicago (which I don't regret for one moment). So, I got involved from the outset, and as William Respess suggested, it is good to get other people whose expertise you need -- accountants or financial advisors if appropriate - at the initial stage, and to start doing things right. People are needed who can be sounding boards and who can play the devil's advocate for the people who are anxious to get their technology commercialized, i.e., the scientists and technicians.

Question: I have a specific question for William Respess which might also be of interest to some other biotech companies. We have recently started exporting and importing. My question to you would be: have you succeeded in getting customs to provide clear-cut classifications.

Answer: Respess: We have not been importing biological materials as such. We have attempted, and are continuing to work at getting some foreign origin cell lines into the United States, but in the one instance I'm thinking about, we've had more difficulty in dealing with the potential licensor in reaching an agreement than we have with the customs people. The Agriculture Department, has regulations which are applicable to importation of Hybridomas(25) and so forth. At this moment, I don't know what they are. The export regulations, of course, are not set forth by the customs people at all. They come from another part of the Commerce Department.(26) The reason Customs gets involved is because Customs is on the scene at every airport, or every dock if you export things like this by ship. And so they become the police officers of the export regulations. And because they're not charged primarily with this, they err on the side of grabbing something, rather than letting it through. You can't be sure if it goes through once, it'll ever get through again, because they don't necessarily look at everything. And, as I said before, it's a doomsday situation once they've got it. They don't let go. I think, in that respect, there are a lot of reasons to look into export regulations before you export, because they're very complex, and detailed. For example, I once saw proteins on a list of prohibited chemicals. That's just about everything, from my point of view, for a company like Hybritech. Fortunately, the prohibitions on proteins are not very stringent -- you can ship them everywhere, including the Soviet Union, and other permissible communist countries. The problem is that the technology that might go with that shipment has a different classification. So while you might be able to ship an antibody-based product with a protein-based in kind, you can't necessarily transport the know-how to do it, and this also includes shipping it to a friendly country, if you know their business is to transport to the Soviet Union, for example. I think the best I can say is that there are many lawyers in Washington who make their living at this and who know the people at the Commerce Department. They can get clearances in a couple of days which would take you six months negotiating through the bureaucratic maze. So, my advice would be to go to one of these lawyers and seek their assistance. The big unknown is the Department of Defense's position on these things. I think all bacteria are prohibited, except from an approved list, and I don't think E. Coli is on that list.

Question for Dr. Pinon: What ethical considerations should a professor bring to the decision of whether to join a commercial enterprise or to form one?

Answer: Pinon: That's a very difficult question. I am trying to put myself in that position, never having been there before. I think that for most of a very large range of possible applications that one could conceive, let's say recombinant DNA technology or procedures, there wouldn't be any real ethical considerations. For most of us, the products that one would be interested in making are products which have some utility, i.e., some therapeutic utility, or which go towards improving plant stocks and a variety of microorganisms. I think the ethical issues come more from the third aspect of the questions that we will be discussing later on in the day -- genetic engineering directed at animals or humans. There is another issue which has to do with the deliberate release of R-DNA or organisms with R-DNA into the environment. This is a very hot topic, and such experiments are generally prohibited except under strict NIH guidelines.(27) Even in those situations, I think the potential benefits far outweigh the conjectural problems. In those cases, I personally have no ethical qualms.

Question: The ethical terms of taking something developed in a lab in large part funded by public money, and developing it commerciallyregarding that aspect of the ethical question, how does a professor approach this issue?

Answer: Pinon: Let me try to assess what I see as happening. I think it is justified - it is clear that this has been going on. The use by a biotechnology enterprise of procedures and knowledge which are developed in the laboratory for private gain is not new. This has been going on for the past fifty years in chemistry and physics and other areas. This in fact, has been done routinely over the years. I don't think those kinds of issues were raised for chemistry and for physical engineers. Certainly, they weren't raised to the same level that they are being raised now with biotechnology. I'm not sure I really understand why. Nevertheless, with respect to the NIH, its policy, as I understand it, regarding this type of application, is one which, while not necessarily encouraging it, certainly does not discourage it. For example, I have an NIH grant, and NIH has primary decision whether I may apply for a patent. I have to ask them: do I have your permission to do so, and routinely, their position has been to relinquish their rights. Next, I have to approach the State University of California, because I am their employee. So, the policy of NIH, as far as I understand it, has been that these discoveries, or the fruits thereof in the long run are going to benefit the public. And if people want to try to exploit them in a way that can't be done at the University, then it only makes sense to allow them to do so.

Comment [from the audience]: I just want to make one comment on that. Seeing it from the other side, many of our technical people were with universities, where they developed the technology, came over to the company, and helped to further develop that technology. We on the other side, in terms of ethics and doing the right thing, take a very strict position. We do an investigation, we see where it was originally developed, and if it appears that there was any connection with the university, we either ask for a disclaimer, or we work out a license. In other words, we try to find out if somebody out there or some institution out there has a proprietary interest in what the professor or the scientist who used to work at the university brings to the company. From an ethical point of view, I don't think any credible outfit (you do have people who get upset "hey, we don't have to show this to Professor So and So - I did that a couple years ago - this is different from what I did there") would not take this strict approach. We do our own investigation, and if it's disclosed that somebody else is also an inventor or it was done in a laboratory, we do the difficult thing, but the right thing. In terms of ethics, using something that belongs to somebody else, is intolerable.

Part Two

CURRENT ISSUES

Chapter Four
HUMAN SOMATIC CELL GENE THERAPY
Dr. Ted Friedmann

You are all aware that there is a revolution going on in biotechnology, otherwise you would not be here and would not be in the business you're in. I think it is clear, that underlying the revolution, at least from our point of view, is the impetus to understand and manipulate human disease.

What I would like to do, and I think what my task is, is to briefly remind you of something about the medical and scientific burdens in the realm of human disease. I will stress certain genetic diseases, and I hope that this will make sense by the end of this presentation. I will discuss the nature of some of the scientific and medical directions, and then will review briefly some of the public policy, and scientific and ethical problems and dilemmas that we're getting into.

I would like to suggest that much of the pressure in biotechnology is, or ought to be, in the direction of understanding the nature of genetics and how it is expressed in human disease. It is certainly at the heart of much of the technology that we have available to us now. I would like you to believe at the end of this presentation, that the technology should be looked at from a long range view -- promises, techniques, procedures, and approaches to diagnosis and therapy which dwarf what we have available today, in the way we approach, understand and treat disease.

The speed at which this field is moving is quite impressive. Perhaps half of the people in this room were born at a time when we did not know what molecule stored and expressed the human genetic information, or any genetic information, for that matter. We did not know whether there was encoding of DNA. We did not know how many chromosomes the human being had. We certainly did not know anything in detail about the mechanisms by which that genetic information flowed from repository molecules to gene products. It has just been within the past twenty years or so that all of this has exploded from under us. Anyone who thinks that explosion is going to slow down in its implications for diagnosis and for treatment is mistaken. If anything, it is going to increase.

Very briefly, we know that genes and genetics have a powerful role to play in human disease. The number of diseases that we now recognize to be genetic is growing; exploding just as the field in general is exploding. We know now, for instance, that human cancer, in all likelihood, should be considered just as genetic a disease as cystic fibrosis or sickle cell anemia. It is most likely always associated with some rearrangement or the expression of rearrangement of genes, or the expression of genes in an abnormal or inappropriate time or in an inappropriate way. So, the understanding of a major public health problem like cancer, or cardiovascular disease, aspects of aging, aspects of degenerative disease, will be associated clearly with defects in the expression of genetic information. Therefore, they are genetic diseases and ought to be approached in that light.

What we know about the human gene and its relationship to disease is that it consists of about three billion bases strung along a linear DNA molecule. We know that a mistake at one position, one mutation, one base in the wrong place, will lead to devastating disease. Let me show you why.

Let me illustrate with the example of someone who is suffering from a disease that is inherited, like hemophilia, and which is called Lesch-Nyhan disease. It results from the absence of an enzyme which is involved in the synthesis of DNA and RNA. The victim is retarded, and has kidney stones which threaten his life and which will kill him. He has severe cerebral palsy and displays strange behavior that involves biting off his fingertips, not his fingernails but his fingertips, and biting off the edges of his lips. This is a genetic disease which results, predictably, in all these symptoms and signs, and the cause of this is that one base out of three billion somehow got into the wrong position. There are numerous examples of these kinds of diseases which I need not stress here. But we are now able to recognize and characterize them -- we are now able in a powerful sort of way to detect them at all stages of development, prenatally and otherwise. And this is an approach that is clearly going to pay off in diagnostic terms, not only for this disease, which is extremely rare, but for many other diseases which are major public health and world-wide health problems. Infectious diseases, in fact, have genetic components.

What we ought to begin thinking about, is that in all these diseases there is a genetic component. We ought to understand the nature of that abnormal component in order to be able to treat this disease well. This (Lesch-Nyhan) is not a treatable disease, and like most other genetic diseases, is screaming for a new approach. I'd like to remind you that we know about 4,000 reasonably well characterized genetic diseases. Let me suggest to you that if you for a few moments thought about the number of diseases that are well treated; noninfectious diseases, nonsurgical disorders, that are well-treated at the moment by medicine, there is only a small handful. Only four or five of these well-developed, well-recognized genetic diseases are reasonably well-treated, and then not perfectly, mind you. The rest of the 4000 are almost in the category of untreatable diseases. That is not a terribly good record for an institution like scientific medicine, which claims to be very sophisticated and highly technological. There is this screaming need for new kinds of approaches to disease.

This is what genetic disease is all about. It begins with the double strand of DNA. The genetic information is stored in that molecule. It is then transferred and transformed into a functional enzyme or other protein. That enzyme mediates some metabolic step, in this case, the conversion of B to C. And, if that enzyme is defective because of a mutation of DNA, this step will not work. When that step does not work, there is too little of this product made. Let us say that that product is growth hormone. If you do not make enough growth hormone, you do not grow very well. If you do not make enough insulin, you do not handle sugars very well. So, the scheme is that the defect in this pathway will interfere with metabolism. The way to treat this kind of disease is to provide this product. So you provide growth hormone, and you give insulin, and these kinds of treatments are reasonably effective. But they are not perfect, if you think about how inadequate, in fact, is the treatment of diabetes with insulin, and the fact that complication incidence is almost unchanged in many cases. This tells you something about the effectiveness of this approach.

There are other ways a genetically-produced metabolic malfunction produces disease. If there's a block in the pathway, the product B, will accumulate at toxic levels in the cells and injure them. We treat this kind of disease by removing B. So you devise pharmacologic agents, drugs and things, which will dissolve this material out of cells out of storage inside the cell. We are familiar with the role of biotechnology in the design and the production of new biomedically active agents, such as enzymes and drugs and so on. This will continue to be a major approach to treatment, and design of drugs that are aimed to remove toxic accumulated [metabolites] will continue to be a major factor in biotechnology. We can also treat this defect and the diseases that result from it by interfering with the input into this pathway, so that if this B is accumulating because this pathway is being fed too much A, perhaps you can eliminate A by dietary therapy, (as is done in Phenylketonuria, or PKU). You thus interfere with the amount of material that goes into the pathway, and therefore reduce the amount of B that accumulates at toxic levels.

It is possible to treat disease at the level of the enzymes -- if there is an enzyme that is missing, or a hormone, or insulin, or some other metabolically active chemical, for example. If there is an enzyme that is responsible for that reaction that is missing, it could be provided in large amounts by purifying the enzyme and simply giving it to people in one form or another. Or cofactors that the enzyme requires for activity can be provided. Enzymes sometimes have to interact with small molecules in order to be active. A defect in this enzyme can either destroy its activity per se, or it can interfere with the way it interacts with the small molecule. Flooding the system with lots of the small molecule should therefore force the enzyme to function. This kind of treatment does work, occasionally. One can replace organs that are damaged. We are all aware of the current interest in organ transplantation, and the likelihood and the inevitability is that organ transplantation will assume a more prominent role in treatment.

But if you ponder at this scheme for a moment, the gene to enzyme to protein route, I think a logical inconsistency will become evident. That is that all treatments that we have now, every treatment that is available for disease of any sort today, is aimed at a site other than the site of the defect. The defects are sitting there in the DNA, they are expressed in mutants of normal function, and yet all of our attention is devoted to treatment distal to the site of the defect. That is understandable, because until recently, that's the only thing that we had available. We had no way of getting to the genes, characterizing them, and knowing what to do about them.

The revolution in genetics, I'd like to suggest, is likely to change that. It is likely to change the way we look at many diseases and certainly, the diseases of genetic components. The suggestion is, of course, that we begin to think about what we can do about the mutant gene itself.

Now, what can be done about mutant genes? One can change the way they function, they can be moved from one place to another in the DNA molecule, or one can try to devise pharmacologic agents, drugs, that will force them to work properly. One can also envision simply adding a new gene where either one does not exist or where one exists only in a mutant form. All of these things will happen. We have already developed procedures for turning on genes. And we all know that genes are not all working at the same time in all of us. Our skin cells don't make blood proteins. Our brain cells do not make liver proteins. Genes are regulated, turned on and off, by mechanisms which we are only just beginning to understand. Drugs are becoming available which have the effect of turning on genes that have been turned off in a particular organ. The first example of this is a drug, a highly toxic drug used in cancer chemotheraphy, which has the interesting property of forcing the expression of the blood protein -- a globin protein, that is ordinarily turned off after fetal life. People who have sickle-cell anemia need some extra globin proteins. They can, at least preliminarily now, be treated with this drug which forces open the expression of this fetal protein, and therefore reconstitutes part of their hemoglobin synthesizing capability.

Another approach would be to add genes where they don't exist. Biotechnology, as it appears now, certainly can take care of this. New drugs will be developed -- new ways of producing highly active and efficiently functioning hormones and proteins and enzymes. All that will come, and it is all necessary. It seems clear that techniques will be developed that will also enable us to manipulate at the genetic level, insert genes, i.e., modifying the action of resident genes -- turning them on and off at will, or putting them in when they don't function to begin with.

I will just briefly summarize one way in which this has been approached by a number of investigators. There is a group of beasts in nature that know to do precisely this job - that is, take a gene into a cell and cause it to work. This is a class of agents known as viruses, tumor viruses, or other kinds of viruses. They have been selected evolutionarily over billions of years to do this job, and the tack that some people (armed with the ability to isolate genes and, in fact, now to synthesize genes), have taken, is to devise a scheme of piggybacking a gene onto a virus of one sort or another. The virus is disarmed so that it doesn't do its own nasty work, but simply works as a vehicle to insert a gene, at the will and direction of the manipulator, into a cell to restore a function that is not presently in that cell.

Mammary tumor virus is a glob of protein inside of which is a small piece of RNA (instead of DNA). That RNA can be replaced by RNA which you would like that virus to carry. It involves complicated, but now rather straightforward methods. You can structure virus-like particles, and there are some other viruses that can be used, of all classes, none of which normally affect humans. You can stuff them with information you would like, and then ask them to carry that information into a cell. That, in fact, can be done, and model systems have been developed and shown to work.

In general, this virus has the following structure. There are two regions that regulate the expression of the genes by that virus. Inside -- between those two regions is a strand of DNA -- or RNA, but in this case, DNA. These two black boxes regulate the expression of whatever is between them, and between them here is a series of three viral genes. Now, you can remove those viral genes, you can replace them with genes which you have isolated or synthesized, retain the regulatory signals, and insert the modified virus into the cells. This material, now, is no longer capable of carrying out a virus infection, because it doesn't have the viral genes to do its own dirty job, but it carries the other genes you would like it to carry into the cells. This gene can be any gene that you have in the laboratory, and, in fact, the gene that is responsible for the Lesch-Nyhan disease has been inserted into this kind of vector and then inserted into cells. In the laboratory culture, those cells have been cured of their biochemical defect. Model systems are now being developed to deliver those cured laboratory cells into whole animals and the scheme for doing that is as follows.

This is the attempt to put that gene, now in its new form in the virus-like particle, into a whole animal -- in this case, a mouse. The idea is to take bone marrow cells from the mouse, inject them with this virus-like agent, put them back into the mouse in the standard kind of bone marrow transplantation experiment, and to irradiate that mouse before you put these cells back. So, if that mouse is to survive, it has to pick up some of these cells you've manipulated by infecting them with the new virus. This forces the mouse to express the new gene that you put into it. Of the mice that survive that nasty experiment, is that new gene working in the mouse tissue? The answer is yes, because one can insert these genes into mice; and as a model system for inserting genes into animals in general, this does show that that technique is feasible. It is not efficient, but it is feasible. The way to prove that this job has been done effectively, is to take the marrow from the surviving mouse, or its spleen and examine it biochemically to see if this new enzyme is present -- and this proves that it is present.

That, in a nutshell, is one form of genetic manipulation which involves the addition of genes to defective gene-mutant systems. There are many, many problems with this kind of technique. It is still certainly at the very early model stage. And yet, because of the ethical, medical, scientific and public policy questions and dilemmas it raises, a variety of agencies have begun to look at this technology and related technologies, and are asking under what conditions will this sort of approach to manipulation, viz., treatment and not necessarily detection, employing these theoretically very simple, easily accessible and equitably distributable techniques, be used?

What ought we know about these techniques before they are allowed to be used on a broad scale? The federal government has gotten into the picture through the Recombinant Advisory Committee of the National Institute of Health, and the Office of Technology Assessment, and the FDA is about to examine the issue as well. The kinds of recommendations being made, given the dilemmas that are posed by these techniques, are typified by a series of questions and guidelines that were published recently in the Federal Register by the Recombinant Advisory Committee, looking at the technology, feasibility and desirability of this kind of approach to disease management.(28) Many of the questions are those that are intended to direct investigators technologically and otherwise in the design of experiments. Most of the questions they are forcing investigators to answer are technical questions - how to design the system; how to test its safety and efficacy. They range all the way from describing the gene and/or virus, to what will be done, if anything to the germ cells, or whether these manipulations involve only body cells which will not be inheritable.

Another series of questions pose societal and public policy questions, which first involve such issues as: what are the eugenic implications of these techniques? What assurances do you have about safety and efficacy? What about the availability of the techniques to insure equitable distribution? The question of the germ-line changes is of primary consideration to the regulators, and for the time being, they would assume that no germ-line modification, i.e., modification of sperm or eggs for the purpose of long term, inheritable change for the cure of disease, ought to be allowed. They have been particularly anxious about issues of animal experimentation -- whether primate testing ought to be allowed, and at what point. This is all in the Recombinant Advisory Committee's guidelines to the FDA. It has not yet been decided what its role should be in the development of this technology. Presumably, they will come up with their own slightly different, perhaps largely different, approaches to the dilemma.

A feature of this kind of work is that it has incredibly high visibility. It has been done with more oversight and with more supervision and control than almost any other scientific technology in recent memory. The reasons for that are readily apparent. There is something viscerally disturbing about some of the techniques. Not concerning the stated aims as I have described them, perhaps, but with the conception that the techniques which are involved in allowing one to do this kind of work, are very close or identical to techniques that one can imagine to be frivolous or mischievous in nature.

In summary, I would like to suggest that a genetic approach to genetic disease makes sense; that it is inevitable, is scientifically and logically consistent, and is in fact, necessary. Again, I stress the record that we have had in the treatment of disease is not a very sterling one. If we can treat well only five or six diseases out of 4000, that's not very good. To my mind, it is likely that absent this approach, the pharmacological approach -- the design of bigger and better mousetraps, the design of different kinds of drugs -- while it will have a major impact on some diseases, will not begin

to approach all genetic disease. The techniques are
so powerful, and so versatile, that a direct attack
on the mutant genes will be possible and ought to be
done. I would like to suggest that the implications
for all this are long term, that the role of industry
and technology ought to have the long term view of
this and related techniques. Biotechnology should
not necessarily limit itself to thinking about design
of new drugs or enzymes, which is aimed at a
peripheral side of disease treatment, but rather to
begin to think about aiming the arrows of therapy at
the target. That target may now clearly be the gene
itself.

Chapter Five
THE BUSINESS PERSON AND THE UNIVERSITY
Ray McKewon

I speak from the perspective of a businessman, since that is what I am. Therefore what I say will not necessarily be consistent with "standard university policy." And, although I am not usually heretical in my views, I will probably be this side of bland.

I present these views and comments based on experiences that I have had negotiating with universities, and also with certain non-profit research institutes, to acquire access to a technology which was perceived to be valuable or to have significant potential utility in the marketplace.

A university is generally considered to be a repository of knowledge. Clearly, it is a place where the quest for new knowledge takes place. A recent historian argued that one of the reasons a university makes such a successful repository of knowledge, was that so few students took very much of it with them upon graduation. No doubt this could be debated.

It is fair to say that a university's mission is teaching. Typically, it is not the commercialization of a technology. Clearly, it is more properly the mission of business to produce profits from the sale of worthwhile products and services. Some of those products or service concepts are, in fact, derived from discoveries made at universities. Now, some social commentators argue that charitable giving and charity itself is a proper objective of business. And really, you will not find too many businessmen who will argue that that is not so. But, as a responsible corporate citizen, although charity is an important thing for them to consider, it generally is a by-product of being successful in business, and in fact, producing a profit. One of the ways they might aspire to do that is to gain the use of a valuable new technology. And, of course, quite often that value was, in fact, created in a prestigious university or institution.

As a result, one encounters a great variety of relationships between the business world and universities. Clearly, there is an enormous appetite in the business world for some of that which the university is able to produce, and produce uniquely. The sorts of things that a university is likely to do with its resources, which include not just money, but the knowledge and talent I referred to earlier, aren't likely to have been produced in business, or be producible by business.

Some relationships between business and universities are non-controversial and of long standing. Universities can, well beyond the traditional teaching of students, attempt to transfer some knowledge and some skills into the business world, and, of course, you've seen, especially in this town, some very effective extension programs that do that. And, you sometimes also see the reverse - businessmen serving as advisors to universities, to help them understand that which it is their business to know. You see intern programs that are particularly worthwhile and functional. And, of course, you see universities providing generally a kind of cultural and intellectual stimulation which makes our society varied and interesting.

Universities, of course, also look for charitable donations with no strings attached. And, that sometimes occurs. Clearly, the major source of capital is the government, but in a time when that source of capital is becoming more frugally managed, alternative sources of capital become important. In fact, capital itself is very important to a university's ability to do all that it sets out to do.

So, if it can't get a charitable gift, or a grant from the government, what then? You heard some discussion earlier today about other possibilities. Corporations, from time to time, do make grants to individual professors or to departments or to schools that are perceived to be particularly proficient in a particular art or technology. Rarely are those grants made without certain strings attached. Typically, the professorial team, if you will, believes itself to be particularly accomplished in technology A. And a company agrees and says "we think you stand the likelihood of producing something worthwhile, and we agree that an appropriate grant should be X number of hundreds of thousands of dollars, (sometimes millions) and if, by the way, you discover something, we would like the right of first refusal for its use." Now, that can take many forms. Perhaps it's a patent, a license, or perhaps it's other things. Clearly, however, it is a right for use of some sort.

There are occasions where universities allow themselves to contract to provide services to corporations where those services are particularly needed. Hospitals are examples where laboratory services and the like are sometimes contracted out as though they were in the business, although that is not their primary mission.

Given that a discovery has been made, with or without direct support from the corporation, clearly it is the case that universities often have something that a company wants. The task for the company is that of acquiring access to it. At this point the company meets with a rather unique creature, generally someone on the parent board of a university, who doesn't seem to act, look, or behave anything like a professor. This person is pure business. And his mission is to maximize the university's inflow of cash from that discovery. So, it is almost businessman to businessman.

A transaction may take the form of an outright purchase of the technology. More often, however, it is in the form of a license, and you'll be hearing more about that later in the conference when patents are discussed. The university usually wants to get its own investment back upfront, and to have a long term royalty stream that it can enjoy and use for its own purposes, e.g., financing further pure research, teaching and general capital expenses.

The other sort of relationship that business and universities often have isn't so much institutional as it is person to person. I have rarely encountered a professor who was unwilling to sell his consulting services. And it is a fairly active business. Professors are, in fact, skilled, and have credentials which make them individually marketable to companies who need access to their ability to guide them through a technology hurdle of one sort or another. And while not often spoken to, that consulting business is really quite a big one.

Having said that, it is important to point out that in order for business to succeed and be well-received at such an institution, it has to behave in a responsible fashion, and deal with a series of very interesting, emerging issues. You hear great debates and read a great deal, about some of the following kinds of topics. Should universities, for instance, be engaged in pure research or is it appropriate for them to engage in applied research?

Pure research is very rarely conducted by business. There are exceptions: the Bell Lab is extraordinary, but certainly it is not commonplace. Anything which of course discourages the conduct of pure research would be our loss, although from time to time, it is considered completely appropriate for applied research to be conducted at the university. Certainly when grant money comes from corporations, their hope is that if it is pure research, it has a ready application and this is the very thing which is being induced by the grant.

There is the whole issue of who owns the discovery, and, of course, you heard that earlier discussed. Because I am in the venture capital business, people frequently call upon us for money. We typically see anywhere from two to seven business plan proposals every day, requesting money for various kinds of projects. Very often, proposals from professors are not in the form of a business plan, but in the form of a spec sheet discussing the technology. We are always curious as to how they made the discovery, and what sort of discovery it is and who might own it. We receive many really intriguing phone calls from muffled voices in telephone booths with remarks like: "I don't have any posted office hours at the moment, I'm not in a class, I'm off campus, and this just now came to me. Therefore, I don't think the university has any claim on this at all." And as light-hearted as I made that sound, it happens.

Generally, people in business really don't mind paying a university for its entitlement, and do not look for ways to circumvent their claim of ownership. Because when they do those things, the controversy they stir-up and the litigation they inspire make the whole attempt too troublesome. So, when I get those phone calls, I say, "it's perfectly all right for you to call me from your office and if I like what you have to say, I'll worry about dealing with the university, and I am sure things can be made to work out."

If that all comes about, then, of course, there are other interesting emerging issues concerning what the university is doing at that point. Is it still a repository of knowledge? Is it still questing for knowledge? Or is it now a marketeer trying to commercialize its technology, which in fact it spent money on? After all, it did build the building, provide the environment which encouraged this research, and it did provide that cultural milieu in which professors and scientists interact and inspire themselves to come up with the things which are very important to the world and exciting to them personally. And now the university is going to try to make some dough out of it, and in essence, market it.

I might call them and say I am interested in the technology that Professor Jones has discovered. It is not at all atypical for me to get the response: "well, what will you pay for it," or "let me think about it," and what "let me think about it" means is they will go out and see if they can get any further bids. They will market it to various other pharmaceutical firms and, in essence, try to maximize the return they have made on what they perceive to be their investment. I am not offended by that -- unless I don't get the deal. But that is the way interactions take place and it is not inappropriate, after all. They are entitled to something for what they did, and if they don't seek those entitlements, they don't get the money they might otherwise have. Without that money, they can't do the things they are so good at doing, and those things really won't get done in any other environment.

There is another interesting issue, and it happens both when you make grants to universities and when professors discover things (and we'll hear this discussion later when patents are discussed). In order to commercialize something, we in business need to have certain claims of exclusivity on a hot topic, or certainly secrecy for some period to succeed at getting to the marketplace in a competitive manner. Absent an ability to give ourselves some proprietary protection, how can we possibly justify spending between ten and twenty million dollars and waiting seven years to get through FDA regulatory hurdles, when anybody else can spend four hundred thousand dollars and have the same thing, after we have done all that work? We could never hope to make a profit.

All this takes place in an environment where the professor's stature is often measured by how much he published on the topic (which doesn't exactly foster our ability to maintain trade secrets and other proprietary protections). The balancing of all this is very tricky, indeed, and done improperly, regulatory people get angry and, of course, we don't succeed in obtaining the desired long-term relationship with the university.

There is, too, this matter I alluded to earlier about professors and business: should professors be in business or should they be professors? And if they are professors, what are professors supposed to do? Is it only teach, is it only research? Will we see a movie titled "The Ph.D. and the Corporate Seal?" Or will we see them locked up in ivory towers or kept there for fear of what they might do to the business world?

There is a story here in La Jolla. It is really a remarkable story and contains many lessons. A few years ago, a hematologist/oncologist, at the UCSD School of Medicine, while working at the University, learned how to pronounce the words monoclonal antibody sooner than most. He was having lunch at Denny's and sitting across the table was a venture capitalist from San Francisco, Brook Byers of the prestigious venture capital firm of Hunter, Perkins, Caulfield and Byers. Brook liked the way this doctor could pronounce the words monoclonal antibody, and decided that if all this was true, it was, in fact, worthy of a modest capital infusion. Some of this comprised a grant, and some went to capitalizing a facility into which some additional talent could be poured.

That company, of course, is Hybritech, and the market value of this company became well over several hundred million dollars. It is one of the first companies to succeed in transferring biotechnology into the marketplace. And it is making a significant and important contribution. Over a hundred million dollars have been invested in that company in the last few years to sponsor research in cancer therapeutics and cancer diagnostics and the diagnosis of other important diseases. Perhaps none of this would have taken place had it not been for some of the early work done at UCSD, and the Medical Research Council at Cambridge, where that technology in part was originally discovered.

So important, though tremendously agonizing issues about the role of the professor arose as the hybritech story began to be told. Should a professor be a corporate officer? Should he be a director of a company? At what point was he thinking his thoughts? Was he thinking his thoughts when he was on NIH time? Was he thinking his thoughts when he was on state time? Was he thinking his thoughts when he was on corporate-grant time? Who owns his thoughts? Can anybody own his thoughts? The Hybritech story got worked out and everybody seems to be very pleased with the result -- because a partner relationship was created between the university and the company.

The last topic, I will discuss is the issue of money. Some of this is caused by an increased appetite on the part of universities for capital. I mentioned earlier that obtaining grant money is more difficult today. Grant applications generally are rated, and even those rated as being attractive and acceptable often are not funded because of the unavailability of funding. Should that research simply not be financed or should it turn to a private, for-profit source, knowing that there might be strings attached? All of this occurs in an environment where universities themselves, and very often individual professors, have their stature measured by how much money they control. One of the reasons UCSD is prestigious is that in the last ten years, it ranks fifth nationally in total grant support. So whenever the National Science Foundation or the National Institute for Health decided to place a high technology bet, they are as likely to place it on that institution as on any other. That gives it a special mystique and prestige which in turn enables it to attract additional talent, and additional corporate sponsorship, and create that critical mass which produces those ideas which are so important.

These are among the current issues created by these relationships. We must deal with these, because if we do so successfully, I think we will see the cure of several important diseases which have thus far eluded us. I think we will see businesses created, jobs created, and many other worthwhile accomplishments.

Chapter Six
CURRENT ISSUES IN PROPRIETARY RIGHTS TO BIOTECHNOLOGY
Ned Israelsen, Esq.

I'm reminded of the story originally told by Judge Markey about three patent attorneys who were discussing inventions that had been made, and the first said, "You know, I think the most marvelous and serious invention of mankind, or accomplishment of mankind, is landing a man on the moon. Just think of it. We've had a man from our planet travel through space and walk on another planet." The second one said, "Yeah, that's pretty impressive. But when it comes right down to it, it's got to be modern communication. You can go home and sit in your living room tonight and watch an event as it's happening halfway around the world." The third one thought for a minute and said, "Well, gee, the moon, that's pretty good, and communications are certainly impressive, but the most marvelous, mysterious invention of mankind has got to be the thermos bottle." The others looked a little puzzled, and he said, "Well, think about it for a minute. If you put a hot liquid in it, it keeps it hot. Right? Right. And if you put cold liquid in it, it keeps it cold. Right? Right. How does it know?"

For the public, or non-scientific community, reactions to the new biotechnology is much the same. It is indeed a marvelous and mysterious science. Yet what it comes right down to is that the benefits of this new science are dependent on protection of commercial developments. The new biotechnology itself has been of sufficient economic importance that the subject of a protection mechanism, so to speak, for that technology has been a topic only for about ten years.

There has been much effort to fit this new technology into existing frameworks for protecting intellectual property. A suggestion was made, for example, that copyright protection might be appropriate, for a new base-pair sequence that is created, as an original work of authorship. This was not successful. There is currently patent protection available for asexually-reproduced plants, (unfortunately it doesn't cover bacteria), and the Plant Variety Protection Act(29) for sexually-reproduced higher plants that may some day assume greater importance in the field.

For now, however, there are primarily two vehicles for protecting the technology. The first is patent protection itself by means of regular utility patents. A patent is the exclusive right granted by the federal government to an inventor to exclude others from making, using, and selling an invention for a period of seventeen years. The Chakrabarty(30) decision by the Supreme Court in 1980 was a milestone in permitting patenting of genetically-engineered organisms. The issue before the court was whether the Patent Office could declare that a living organism could not be patented because it was a product of nature. The Court held that "living" or "nonliving" is an irrelevant distinction in determining whether something is an article of manufacture, and that the real focus should be on whether the invention, whatever it is, was altered or modified by man into something new. Subsequent to this landmark case, there have been a few hundred biotechnology patents granted for genetic engineering, hybridoma techniques, and gene splicing, and there are over 2,000 filed applications which are pending. Patents of this type can be extensive in scope. Patents may be obtained covering a process, a manufacturing method; a method of using a product; an apparatus, such as a cell-sorting machine; a diagnostic kit; a chemical or biological product, such as a new pharmaceutical; a polypeptide; or even genetic material that has been altered by man, or living organisms which are not naturally occurring. This also extends to pure cultures of naturally occurring organisms where a pure culture does not naturally occur.

Many steps need to be taken to secure valid patent protection. We have talked about employee agreements today, and about the issue of resolving inventorship ahead of time. In order to have a protectable invention, it is important to keep detailed records of its development. Laboratory notebook procedures should be followed in research organizations to help assure this protection.

Publication and disclosure policies are a sore point between biotech researchers (who usually come from an academic background where publication is the rule) and patent attorneys (who are concerned about the loss of rights.) Premature disclosure of the invention may bar a patent on that invention. In the U.S. there is a one-year period from the time of first disclosure, offer of sale or public use of an invention in which to file the patent application. (31) Unfortunately, with respect to obtaining patents in most foreign countries, the application must already be on file in one of the countries (probably the U.S. in your case) before any public disclosure, offer for sale, or any public use of the invention.

A second protective device is that of trade secrets. A trade secret is information that derives independent economic value from not being generally known and which is the subject of reasonable efforts to preserve its secrecy. Trade secrets are short-lived in the university environment for obvious reasons. Unlike patents, they must always be kept secret if they're to be protected at all. Furthermore, the protection goes only to wrongful misappropriation, and not to independent development of the idea by another, or reverse engineering. For example, if your trade secret is a process for making a particular protein, and a person, in studying this protein, ascertains the process by which it was made, that person is free to use that process and perfectly free to use your now-discovered secret. You have no protection.

There are other important considerations in determining whether to rely on trade secret or patent protection. A case-by-case analysis is essential. The decision depends, for example, on the type of research organization involved; the type of invention (i.e., will it be disclosed anyway by marketing it?); the marketing strategy; the potential for reverse engineering; the potential for unauthorized, accidental, or forced disclosure (e.g., in connection with filing foreign patent applications, which are automatically published). There are also the questions of microorganism deposit requirements, and whether there is a desire to license the invention, which is much easier when a patent has been obtained.

The deposit requirement raises some very significant legal issues. The quid pro quo for obtaining a patent is full and complete public disclosure of the invention, including the best way known to you to practice that invention. Where a patent involves genetic materials, or living organisms, we do not generally have the ability to fully describe these materials on paper, or even in drawings. So how are we going to satisfy the full disclosure requirement? The system that has evolved on a worldwide basis is the depositing of the microorganism, plasmid, DNA, or whatever, in a depository, with the agreement (depending on the country) that this information can be accessed by the public, either at the time the patent issues (in the U.S.) or before issuance when the application is published (in the European countries). This creates understandable problems. Bart Rowland, who obtained the landmark Cohen-Boyer patents covering fundamental recombinant DNA technology, likens the deposit of a microorganism in a publicly available depository to turning over to your competitor the key to an operational factory to practice the invention that you have patented. Microorganisms, which are designed to turn out a certain protein, have all the mechanisms built into them necessary to turn out that protein. It may have taken you three years to develop that microorganism. A competitor could obtain that miniature working factory from the depository and be in production within weeks or months. So the deposit requirement is certainly a sore point, and attorneys would logically try to avoid depositing whenever possible. Exactly when and under what circumstances deposit is absolutely necessary, however, remains largely unresolved.

Another problem that is perhaps unique to this new technology is that the working tools, the hammers and scalpels, if you will, used in biotechnology, were all recently invented by someone. That someone in many cases filed a patent application. And in any particular field, chances are that the fundamental technology that everyone is using is subject to that patent. As more patents issue covering fundamental processes and products, the problem of simply keeping track of who needs to be paid royalties on what emerges. And royalty payments rapidly become burdensome. The Cohen-Boyer patent(32) to which I alluded, and which is infringed by virtually everyone, is being licensed aggressively by Stanford.(33) Its licenses cost $10,000 a year, and the royalty rate runs from a half a percent up to 10% of the product's sales. At last count I believe it has 73 licensees.

As a result of its licensing strategy, Stanford has built up a warchest of some several million dollars to defend the patent and to police it. Other Johnny-come-latelies have had more limited success. Columbia owns the patent by Richard Axle which they are licensing for $30,000 a year, with royalties ranging from 3% to 15%. It has 11 licensees to date. On the other hand, Harvard has thus far been unsuccessful in licensing some very exciting fundamental technology developed there by Mark Ptashne. This must be a disappointment, of course, particularly in light of Stanford's success. I see the biotechnology community becoming increasingly cautious and skeptical about taking licenses, at least until the first round of cases come through the courts establishing the validity of the patents.(34)

A second significant problem here is the enforceability of biotechnology patents. To date, not one single genetic engineering patent has been fully litigated, i.e., has been found to be valid and infringed. This in turn reflects the uncertainty surrounding patent validity in this field.(35)

Another question on enforceability is whether research use alone constitutes infringement of a patent. We know that 95% of the use of biotechnology now involves research. The <u>Bolar</u> case(36) came up to the Court of Appeals for the Federal Circuit (which now hears all patent appeals in the pharmaceutical context). Bolar was trying to get the jump on a patent that was expiring -- they were preparing their FDA data just a little ahead of time. In fact, they jumped the gun by three or four months, and started making the patented pharmaceutical. The court said that because of the commercial purpose behind that research, it constituted infringement of the patent. This, notwithstanding the earlier legal doctrine that research use alone does <u>not</u> constitute infringement.

The case was subsequently overruled by legislation which allowed for pharmaceutical inventions where a patent was about to expire.(37) However, the applicability of the <u>Bolar</u> case in other contexts, notably biotechnology, has yet to be determined. Probably research use for commercial purposes does constitute infringement, but we have no case saying so.

Locating infringers and proving infringement makes enforcement of patents difficult, particularly when the infringement is occurring in the secrecy of a private laboratory and involves microorganisms and genetic material.

A final question on enforceability involves something known as the exhaustion doctrine. Now, if a watch were patented, and you owned the patent and sold the watch to me, you would have exhausted your rights in that patented watch. I could then sell that patented watch to someone else without any accountability to you. But, we have never before had self-replicating inventions. Suppose that under a license from Stanford you engineered an organism which cost you 30 to 50 thousand dollars, and you then sold that organism to a large pharmaceutical company. That would be an authorized sale by a licensee under the patent. But suppose the pharmaceutical company in turn replicates and uses that organism. Query: Under the exhaustion doctrine, are they infringing the patent, and do they have to pay royalties on the millions of dollars' worth of sales based on that microorganism? This question remains unresolved.

I now turn to biotechnology litigation. There are a number of cases currently at the initial litigation stage. Here I mean not only litigation in federal court on questions of patent validity, infringement, ownership of patents and trade secrets and their theft, but also actions occurring in the patent offices, such as federal interference proceedings, where two individuals file patent applications on the same invention and opposition proceedings to try to prevent to issuance of patent to some group not entitled to that patent. Currently, Hybritech, Ventrex, Scripps, University of California, Monoclonal Antibodies, Inc., and Genentech, among others, are in some sort of litigation proceedings.

The economics of patent litigation are staggering. A patent trial will typically cost half a million dollars per side. On the other hand, when the products come on-line and you are paying a 1% or a 3% or a 10% royalty on what you sell, the economics become understandable. Since the Chakrabarty(38) decision in 1980, no Court of Appeals has decided any of the tough issues about how biotechnology fits into the existing structure for protection of inventions. And that uncertainty itself perhaps has spawned much of this litigation. There seems little question but that the litigation will increase until some reliable answers are forthcoming.

In summary, biotechnology <u>can</u> be protected, although no new protection mechanisms have emerged. The existing systems are being stretched to fit the new technology, and in many cases, it's a strange fit indeed. The industry's growing pains will, over the next decade, doubtlessly lead to some very significant litigation.

Chapter Seven
CURRENT REGULATORY ISSUES
Vincent Frank, Esq.

I'm going to talk this afternoon on regulatory issues involving the Food and Drug Administration. I intend to give you a primer on some of the major hurdles one must clear in order to get diagnostic products and drugs through the FDA and, for each category, discussed what I feel are some interesting regulatory issues.

Nearly everyone involved in biotech would agree that the FDA is an obstacle in the marketing pathway. You can expedite the process, however, by educating yourself about the FDA itself, prior to product development. Performing critical studies in the planning stages will definitely facilitate the product development stage. The FDA's role is to insure that no misbranded or adulterated foods, drugs, devices or diagnostics enter the marketplace. It exists to insure that what you ingest or use in a medical sense is safe and not defective.

I will mainly discuss two functions of the FDA -- diagnositcs and therapeutics. The Center for Medical Devices and Radiological Health regulates devices and diagnostics. This Center would oversee the development of a biotechnological diagnostic. It will classify the diagnostic into one of three classes. Class One is the easiest classification to fulfill. Class One products involve only <u>general controls</u>. Now the term "general control" itself is a term of art. Hospital bed pans, or machines that analyze various reagents -- e.g., diagnostic instruments may be Class One products. Generally, these are very simple products that do not pose any health hazards -- or significant health risks. Class

Two products are ones which require "performance standards" -- another term of art the FDA uses, since in the nine years of this regulatory regime, it hasn't developed any performance standards for diagnostic products. What is really involved, apparently, is a process you have to go through. Many *in vitro* diagnostic products, now being developed by biotechnology companies, fall into Class Two. Class Two products are ones which could pose a risk to health if the course of action that is suggested by the diagnostic is incorrect or potentially harmful.

A Class Three product is one which involves premarket approval (PMA), and here is where the most interesting issues concerning biotech diagnostics arise. Now with respect to Class Two *in vitro* diagnostics, you can use a procedure which is called a 510(K)(39) submission. Basically you submit a notice to the FDA that you intend to market in ninety days an *in vitro* diagnostic based on its being "substantially equivalent" to an *in vitro* diagnostic which was on the market prior to May 28, 1976, i.e., the date that the 510(K) provision -- then a new regulatory regime for devices and diagnostics -- became effective. Over the years, companies have tried to show "substantial equivalence" by using for comparison tests that were submitted as 510(K)s after 1976, which themselves referenced pre-1976 tests. In other words, by piggybacking to other "substantially equivalent" diagnostics.

Now if you can do this successfully, you've gained a major advantage. If the FDA fails to get back to you within 90 days (which is unlikely), you are thereafter "cleared" to market the product.

The amount of clinical data and testing involved for a 510(K) is generally less than that involved in the Class Three, premarket approval ("PMA") process. Also, the monitoring that is required after premarket approval is obtained is not required with a 510(K) product. Many companies developing brand-new technologies, like monoclonal antibodies or DNA probes, have been creative in showing substantial equivalence. Hybritech, for example, showed that its monoclonal antibody test was substantially equivalent to a pre '76 test even though the technology wasn't around before 1976.

In point of fact, the substantial equivalent doctrine is a convenient fiction. The FDA is trying to expedite submissions and the marketing of products whose safety and efficacy can be demonstrated with generally more data than the traditional 510(K) application, but which can generally be shown to be safe and effective without a PMA. A 510(K) traditionally does not require substantial clinical testing, clinical trial, or the correspondingly heavy costs and tests. But the FDA *is* demanding that you present more data than seems technically required, similar to the data needed in a full premarket process. Thus, the unofficial term for such an expanded 510(K) is "mini-PMA." In other words, you make your submission for clearance to market, but in your 510(K) submission, you present the kind of data that you would have collected for and included in a PMA. Thus, less time is required getting through the FDA than in the full PMA process, and the regulatory environment after the product is on the market will be a little less stringent. One need not, e.g., file supplemental PMAs for every change to the product. Hence, marketing is both faster and easier overall.

The problems involving regulation of therapeutics or drugs are a little different. Compared with diagnostics, the therapeutics or drug approval process involves more steps and data submissions. Animal tests determining efficiency, safety, and potential toxicity are submitted to the FDA as part of an investigational new drug ("IND") application for approval to begin clinical trials on humans. After completion of clinical tests under an IND, a new drug application ("NDA") must be submitted to the FDA. Upon approval, the product may be marketed. Typically the entire process can take three to five years or longer before a product can be marketed. Of course, this presents many problems when your product is based on an issued patent and the 17-year clock has begun to run.(40) This simply cuts short the product's useful commercial life, and thus translates to loss of big dollars. This is one reason why drugs are so expensive.

One of the larger issues surrounding a recombinant DNA (R- DNA), product is whether the product is the same as a product which is currently on the market, but for its manufacturing process. Is this product subject to a NDA from the FDA? For example, if you can manufacture insulin from an R-DNA process or some other process and it's the same end product as derived from more traditional processes, is a full submission necessary? Is this really a new drug or is this just a different manufacturing process for the same drug? The FDA by its actions has decided to take a case-by-case approach in evaluating recombinant DNA drug products.(41) In doing so it looks at factors such as identity to previously approved products, method of administration to the patient, and the clinical indications and experience with the product as produced by conventional means compared with the R-DNA technology. The greater the correlation between these factors, the less protracted the approval process is likely to be. But this is not an official position.(42) The FDA has thus far only issued a number of guidelines and positions on various R-DNA produced products. It has shown, in other words, a willingness to cooperate and communicate with companies using R-DNA technology. The FDA is, in essence, still learning, and it is encouraging dialogue with manufacturers. In the preR & D phase of a product's geneology, it is admitting to a certain flexibility, but the formal requirements are still in place, so that the formal supplemental new drug or abbreviated new drug applications cannot be used to expedite the process.

The Office of Biologic Research and Review of the National Center for Drugs and Biologics recently issued a draft of a paper entitled, "Points to Consider in Production and Testing of New Drugs and Biologicals Produced by R-DNA technology.(43) This provided certain suggestions for controlling safety, purity, and potency of new drugs produced by R-DNA technology. But for companies considering development of R-DNA-derived drugs and biologics, these guidelines are not mandatory. Initial compliance with them, however, would probably facilitate the regulatory process. These guidelines cover such matters as expression systems, master cell banks, production and purification, characterization of products, and other factors involved in testing and producing an R-DNA product. So the FDA, rather than remaining completely silent on the issue, is trying to make the people involved aware of what it expects to see on the applications. One must understand that the FDA started out as basically a cop on the beat for protecting and prosecuting violations -- this in the days when slaughter houses and bakeries operated under very unsanitary conditions. The FDA was formed to enforce health protection. It has now become more of a regulatory agency consisting of many bureaus and arms dealing with a host of issues. It does issue guidelines; but perhaps more importantly, it establishes and encourages dialogue with the business community.

For example, there is the Office of Small Manufacturers Assistance. It wants to help companies which are getting involved in developing diagnostics or therapeutics from new technologies. It makes them aware of what the FDA expects at any given stage of development. So this dialogue is most important. Much of the FDA's concern in this area is with the risk to public health which a particular product or process may pose. In the case of an R-DNA-derived product or other product produced from new technology, the regulatory structure is slow in changing and adapting. This is not due to the fact that the FDA is taking an adversarial approach, but rather because of its incomplete knowledge; and the real and perceived risks of making a mistake are just too high. This is particularly so considering the emotional issues surrounding genetic research and gene-manipulation. But this will surely improve as a function of better communication, increased experience with the new technologies, and hopefully, a deeper understanding by people at the FDA.

QUESTIONS AND ANSWERS

Question: What is the treatment of foreign data with respect to FDA approval -- is it accepted by the FDA, should it be, etc.?

Answer: Frank: Generally, it isn't. Generally, the FDA requires studies which, for example with respect to a diagnostic device, are done in multiple centers in the United States using controlled and uniform testing. You usually won't get an application approved, or a 510(K) submission cleared, if all you have is foreign data. However, you must present foreign data to the extent that you have it -- thus, for a negative reason. Suppose a couple of deaths resulted from the drug's use in Europe -- the drug you're now trying to clear in the United States. That information must be given to the FDA and failing to do so can result in both criminal and civil liability, as Eli Lilly & Co. discovered in the case of Oraflex. So, to the extent that foreign data might cast a negative light on your application, you've got to live with it.

Question: Is there a bottleneck in your particular line or stream, and if there is, what is it, and ought there be a way of unclogging it?

Answers: Friedmann: It depends on what kind of patient. There are bottlenecks to the sense of urgency we feel about pursuing certain kinds of research: some for laudable reasons, some for reasons not quite so laudable. We naturally want to be in there first. But I think in general, most people probably don't feel anymore repressed now by the regulatory requirements, or by the traditional funding problems for teachers, than they always have about these types of impediments to their work. But it seems to me that we have learned a lot in the last decade or so about functioning with imperfect knowledge. Medicine has an urgency to do that and always will and that draws people into clinical situations perhaps more than it ought to at times. A simple answer to your question is yes and no. We feel the urgent need to get on with the technology, but I don't think we feel are particularly put upon by what funds are available, or what's being required of us.

Answer: Israelsen: Well, in my area there have been a lot of bottlenecks which have already been eliminated, as with the Chakrabarty decision. And it's possible now to have the term of your patent extended by up to five years -- for the period involving FDA or another regulatory process.(44) The deposit requirements for microorganisms are, I think, certainly an obstacle. I think they should be modified to restrict public access to those deposits because the public cannot use these without necessarily infringing the patent. I think it's fine for them to be available after the patent expires.

There is an international treaty, the Budapest Convention,(45) that now permits depositing of the microorganism in one country to serve as patent protection in all member countries. But there needs to be a lot more uniformity internationally as to the deposit issue. The various patent offices are taking steps in those directions.

Another problem concerns the U.S. Patent Office itself and its lack of biotechnological expertise. At the time of the Chakrabarty decision, the office didn't have technically trained examiners in microbiology, for the most part. They are now attempting to remedy that situation.

Answer: Frank: If I understand your question correctly, you want to know about obstacles to converting technology into application. And, because of the regulatory environment, I don't see that it's a hurdle. But there is plenty of information available on how to do it right. And if you go into your business, your development, knowing what you have to do, and anticipating some of the problems that are going to come up, then it's going to be a lot easier. The regulatory environment is just another variable in your business plan. Then you ask, to the extent that the regulatory environment represents a given obstacle and must be addressed from the outset -- how do you best deal with it? That seems to be the real question? This must be done at the product research stage so that you are already accustomed to calibrating machinery, keeping records, etc., all of which may later be required for the FDA.

Application of the new technology is itself, a major obstacle. Business people are finding out that embodying these "cutting-edge" technologies in easy to use, useful forms, or even mundane types of things are major problems. Many companies currently in the product development stage are facing these very problems, despite having a technology that works quite admirably in a controlled, laboratory situation.

Answer: McKewon: I may have more bad news than good. There are some obstacles presented to a business man attempting to commercialize an intriguing new biotechnology discovery which are unique to our time. We've talked about research budgets. We ought to talk about the money which is necessary to take that which is produced by a research budget into the marketplace. This cost can be enormous. A couple of years ago, we were in a bullish initial public offering market, and almost anybody with a Ph.D. and an attractive underwriter could get millions of dollars and have their stock go up -- everybody thought there was something terrific happening. Underneath all that, though, there was this attempt to translate that technology into the marketplace. Basically, there were such obstacles as simply finding a building and spending anywhere from $50 to $110 a square foot just to improve it and create a facility necessary to take the product from the research stage through proto-type testing and clinical trials to the marketplace. And this process proved to be a much slower process than had been anticipated. Also, high value had been put on the sexiness, if you will, of the potential application of high technology -- suddenly the stock went from $20 to $30. But when the market finally began to appreciate the time and money required to commercialize technology, the capital available to finance development-stage companies, which have as their purpose, commercializing this sort of technology, significantly diminished. Now, sources of such capital are far more judicious in the choices they make about what kinds of projects to back. They must be much closer to the market than previously. Also, the technology leap must be over a gap perhaps not so wide as they had tried to span before. And mindful of these things, fewer dollars flow into projects which are perhaps in some ways less promising because they are more risky.

Now the tide turns, the pendulum swings, and we'll have another bullish stockmarket. People forget everything they've learned and this will start over again. But we are just not in that particular cycle at the moment.

Question: Rather than focusing on obstacles; what incentives are there or should there be for the activity in which the participants here are involved to bring the technology to the marketplace. What sort of incentive system is needed?

Answer: Friedmann: There is a failure or inability, I think, under some conditions of the private sector to look more than just at arms length at the technology. The federal government is running, as if there were a plague, from funding the kind of work that led to biotechnology as we know it today. Who in the federal government nowadays would think it clever to fund research problems that deal with how bacteria is reproduced and recombined? But it would be very easy to sell in today's research and funding climate compared to 10 or 15 years ago. What I would like to see is an incentive from the business end, as some of venture capital in some form or other, for a long- term research. This is required for innovative research 15 or 20 years from now. This should begin to play a more prominent role in how all of you think about the field. There is in American biotechnology, a myopic view to get in and out of providing products -- such is a limited scope, lack of imagination in this regard is a limiting factor.

Question: Are business bottom lines and the lure of large dollars for researchers to become partners in more immediate, commercial applications, in fact having a significant impact on the willingness of your colleagues to engage in pure research?

Answer: Friedmann: I think that's inevitable. The government, as you well know, announced its intentions to reduce the total number of research grants in the next funding period from 6500 to 5000. That's an incredible reduction in unfettered research support. Consistent with the stated political aims of the administration there are some realms of human activity that don't fall within the interest of the federal government, and one of these is the pursuit of knowledge and research. I don't know which is the aberration -- the current statement or the last 34 years. It's clearly, a denial of the value, utility and the impact of previously funded work.

Answer: Israelsen: As far as incentives, I think a patent is a marvelous incentive to research and development. Without the ability to protect this representation of capital against plagarism, investment capital won't be forthcoming. So anything which improves the patent system will, I think, be a real incentive to R&D. Educating those responsible for the inventions about the patent system and the effects of premature publication would also be desirable. Hopefully this conference today will have served that purpose to some extent.

Now regarding the royalty-stacking problem which exists. Someone involved with the Cohen-Boyer licensing program at Stanford has offered the idea of a patent pool to eliminate this issue. This essentially is a single search you can go through to get your patent licenses for fundamental technology, instead of going all over the country. Possibly antitrust violations might be involved, however, and without a dispensation from the U.S. Attorney this concept might not work. But, as there has been some movement to make it more difficult to receive capital gains treatment for patent royalties, to the extent that this serves as a disincentive, validation of this type of recommendation would be welcome.

Answer: Frank: Biotechnology holds the promise of vastly improving the "human condition." Although perhaps a contradiction in terms, to investors this means big bucks. These are both very significant incentives, I believe. Some of the long-term financial obstacles have already been mentioned. But I think those trying to commercialize the technology should level with the investment public -- it is a long-term process. It's amazing how the venture capital process gets cast in a certain light which suggests that things are going to happen a lot faster than is the actual case. Pressure is created because the investment public wants things to happen fast. I personally don't think this is good, although it is a fact of life. I'd like to see this change. The ultimate payoff in biotech will be quite large -- it will be worth the money and the wait. Finally, with respect to R & D partnerships as a type of financing, I believe they create more problems than they solve. People investing for tax writeoffs finance deals that shouldn't be financed in the first place. When these don't pan out, the whole industry suffers. Bad deals, bad financing, and the "lemming" attitude of many investors during "go-go" periods create problems which in the long run hurt the credible companies. These are serious business people who see biotechnology as a serious business.

Answer: McKewon: On the R & D partnership front, the tax incentive really is no different from that for oil and gas exploration. It serves as an incentive to certain kinds of behavior. We're permitted to deduct drilling costs in searching for oil and gas, and the reason for this is to encourage discovering oil and gas. True, there is now an abundance of gas -- but this was not always so. It was very much in our best interest to encourage, through the tax system, exploration of oil and gas partly because of the business risks involved. The same is true of biotechnology. R & D partnerships have been an extremely vital financing technique, providing amounts of capital which would not have come from any other source. Genentech has gotten well over fifty million, perhaps (I'd have to review the data) as much as one hundred million dollars in that form. Hybritech has similarly received over eighty million dollars. And other biotechnology companies have used this technique as well.

Private investors would otherwise return those dollars to Uncle Sam's general account, if you will, instead of giving those tax dollars to a specific company for a specific purpose. Some of those purposes will not be worthwhile -- very many will be. But clearly the payoff in biotechnology, the potential payoff, makes most of the obstacles that have been described worthwhile confronting. You have to leap over the financial hurdles, and be patient despite the current initial public offering market not rewarding your efforts. You must consider alternative sources of financing besides NIH money or general funds of a university. You must do whatever it takes to wait-out the commercialization of extraordinary discoveries. I once met a scientist whose claim to fame was that he kept Smith-Kline from cancelling the Tagumet program. Smith-Kline could not understand why this compound would be worthwhile. Over $750 million dollars worth of Tagumet was sold last year.

I met another fellow whose friend was the scientist who in the laboratory licked his fingers while experimenting with a dipeptide. (When he licked his fingers, he said, "this is sweet." And that was aspertane, "Nutrasweet.") And very many times G.D. Searle said, "why are we spending so much on something that merely tastes sweet?" Yes, it does take an extraordinary amount of patience to get through all of this! Generally speaking, it takes a champion who sees it as a personal cause to overcome all those obstacles. Without this kind of personal conviction. These things do not generally occur. And in addition to all of this, it takes a lot of money to succeed, and if a small company or individual professor doesn't have it, they can't do it. And patience is of no account here.

Question: Speaking to the issues of competition and secrecy, as a member of several boards of directors do you find yourself in a conflict of interest situation?

Answer: McKewon: That's a particularly insightful question, and certainly troubles me almost as much as it does you. In fact, I'm no longer chairman of the board of the Immunetech. One of the reasons for this is that I became more active with Cytotech and became its chairman. And given the fact that both were eagerly exploring various approaches to autoimmune diseases, it became clear, regardless of efforts to the contrary, that there is such a thing as too much knowledge, and that I might inadvertently spill this knowledge from one source into another. Eventually, in the long term, a resolution might have to be adopted and I'll have to resign from one board altogether if the technology got so close that a conflict might become real and apparent. Now in the venture capital business, it is a common occurrence for technology to eventually overlap. This is not always so at the inception -- the technology may appear to be very narrow. But these are very clever guys, and they find all kinds of ways to apply that narrow technology and overlaps are created. So all kinds of internal controls are created. Secrecy agreements get signed; board resignations sometimes are called for to avoid what then become real conflicts of interest. And when that's not done, the result may be cloudy registration of securities. And you try to "Blue Sky" those securities in some states;(46) and when this is not possible, you can't sell the stock, and you can't be paid for all your efforts. So your question is even more insightful than you may realize. It's a real concern in our business and we try very hard to keep from shooting ourselves in the foot.

Comment, Bohrer: I learned something about the way we all work together and about how the technology develops from this last panel. If we are going to talk about current issues, we need to do so in our role as problem solvers. So I thank Dr. Friedmann, and Ned Israelsen, Vince Frank and Ray McKewon for being so candid with us about the uncertainties that exist in their lives at this point in time, as they grapple with the continuing problems that this technology presents.

Part Three

FUTURE TRENDS

INTRODUCTION: PERSPECTIVES ON THE FUTURE
OF BIOTECHNOLOGY
Robert A. Bohrer

Having just spent considerable time wading through some of the current issues, in this last panel we now shift our focus to those future problems we think we will encounter. To do so, we must use our experience as problem solvers to construct, as best we can, models based on past experience, that will enable us to deal effectively with the future. We have as our science instructor, Dr. Clifford Grobstein of UCSD and representing business, Mr. Harry Casari of Arthur Young. I will then discuss the future role of lawyers as regulators.

Chapter Eight
FUTURE CONCERNS FOR THE SCIENTIST
Dr. Clifford Grobstein

I will now talk about things that we know very little about, namely the future. Hopefully it will be easier to resolve among us what the future might look like, given the preceding discussion of the past and present perspectives. In fact, I'm not a futurist, either in terms of background discipline and experience or in terms of present focus. Nonetheless, I do recognize that our notions about the future play important roles in what we do in the present. So I think it is worth considering how people are viewing the future development of the technology and its prospects, in order to deal more adequately with the present issues. I guess I can claim some experience in dealing with the things that happen to complex systems when time passes. Being an embryologist and developmental biologist, my business, so to speak, is to see what the effects of time are on complex living systems. And I have had the opportunity to follow this particular system over time and perhaps, I can therefore say something about where we may be going and what sorts of future decisions may be necessary.

I recall that one decade has passed since the recombinant DNA controversy peaked in terms of public intensity. 1975 was a time of great controversy and debate about the significance of recombinant DNA. It seemed that in the mid- 70s, the debate was over how much of what was going to happen would be good news and how much would be bad -- how much represented new opportunity, and how much represented imagined dire prospects of various sorts. Now, looking back (and after all NOW was the future then; in fact, the controversy was expressed frequently in terms of what would things look like 10 years from THEN -- which is now), I think it's worthwhile to look at how things are today in terms of the disagreements, the controversy, and countless projections that were made 10 years ago. I think it's fair say that the reality of today is neither as bright as the most optimistic predictions nor as dire as the most pessimistic predictions of 10 years ago. We are somewhere between. The name that I use in referring to this current period is the "second wave." The second wave is thus not exactly as it was projected at the time of the first wave, and probably the third wave, maybe a decade from now, will not be exactly as we projected. But the most reliable thing one can do with respect to the future, is to recognize that the present always contains some of the past, and if that continues to be the case -- barring the occurrence of some great perturbation, then the future is likely to contain some of the present. Thus obviously, the decisions that we are now making will effect what the third wave will look like. So I view this assignment, from the perspective of one considering the current scene, as a launch pad for future prospects in relation to how it looked ten years ago.

First of all, I think it's clear that the most feared consequences ten years ago, namely biohazards -- the possibility that manipulation of genes would create new organisms, and which due to inadvertent release, would threaten health or the environment -- I think it's fair to say that those fears have so far not materialized. There is to my knowledge no single instance in which any harmful effect has been traced to any recombinant process originating in the laboratories or industrial centers. There is no claim that any such things occurred. Now is it true that we have had over the past ten years constraints, confinements and restrictions that may have contributed to this fact. One might suppose that absent these, maybe something would have happened. But in fact, I think that most of the involved people have reached the conclusion that biohazards resulting from inadvertency are of a very low probability. The circumstances required, for example, for a new organism to become viciously pathogenic, are far more complex than what is done in recombinant laboratories in general. Most feel that one would have to devote oneself to that as a problem; to create such an organism rather than to have it happen as an inadvertency. So the biohazard issue in that original form has not realized. And today, while it is not regarded as something to be ignored, it certainly is not something regarded as so intrinsically risky as to threaten the likelihood or the desirability of pursuing this line of research.

I think it's important though to note that while concerns over inadvertency have been reduced, they haven't disappeared. Concerns over specific applications of these techniques have not decreased. The question of deliberate environmental release is very much a current issue, much more an issue for early decisions than 10 years ago.(47) This is obviously being worked through at the present time, and legal maneuvers essentially have either halted or substantially slowed movement in this direction.(48) This is not a moratorium, but it's a period not too different in terms of environmental releases than 10 years ago respecting the go-ahead with recombinant DNA. The question is, under what circumstances should one go ahead? Under what circumstances can one conduct releasing trials in order to find out what kinds of consequences to expect. In this area, we still have biohazard uncertainty as to environmental and health effects, and there will be the continued need for attention.

The same can be said of biological warfare. It has, from the beginning, been noted that conceivably these techniques might deliberately be used to produce militarily useful weapons.(49) To many, however, it is not a high probability, because again, it doesn't seem likely to be enormously successful. Nonetheless, we are seeing suspicious building toward increasing activity on the military side (according to the Department of Defense) and in response to indications of the Soviet Union's interest in biological warfare. I think most people who have looked into this know that there is no separation to be made between defense and offense in the area of biological warfare. Anything that is defensive increases the possible success of offensive weapons on the other side, and therefore calls for a response of offensive development on this side. So we do have a biohazard problem here, and one which I think bears watching.

We also have concerns about what is now being done with cancer virus and oncogenes. This again is an area that has only been opened recently, and it isn't clear exactly what kind of risks may be involved. It was an area of concern 10 years ago, and remains, in terms of its immediacy, still unresolved. There are thus, continuing, legitimate biohazard concerns. Now again, looking back 10 years ago, I think it's clear that many of the expectations that recombinant DNA would lead to major new scientific advances and understanding in the area of genetics and development, pathology, etc. have obtained. And it also seems clear that these advances will continue for a considerable period of time, certainly at least into the next century, and that they will continue to spawn new technological opportunities. This is assuming that we maintain the level of effort that we have over the last three decades. There was some discussion earlier about budgetary crisis which threaten these continued efforts and ethical considerations concerning interactions between universities and industry -- between the academic concerns with expanding knowledge and the corporate concerns with profit making. These are in many ways two different worlds; yet they have to work together. Because what we are talking about, leaving the economics out, is how knowledge is translated into human benefits. Universities will not carry out development and marketing, and corporations will not carry on basic research support at the level of the federal government. That's part of the problem of relating the two. There is now, as never before, a need for a system of direct feedback from the value generated by research to the support of new, fundamental research. This feedback mechanism is very important. This is a time when we are concerned with application, and the support for basic research consequently has declined, because it comes out of the general welfare budget. When that budget gets cut, research gets cut. The motivation for that is entirely different from the one that is simultaneously operating to effectuate the efforts of 10, and 20, and 30 years ago in the commercial sphere. There needs to be a closing of the loop. There has to be some kind of feedback. And I think that in consideration of the interaction between academics and corporate activity, this should be of high priority; not just to insure that each gets what it needs, but that each cooperates with the other, this being necessary for progress in this area.

Now with respect to expectations that recombinant DNA would lead to practical applications (and this was certainly expected 10 years ago), it is clear that these were justified. However, it is not clear yet that the predicted commercial success will occur. It has certainly not been demonstrated on a large scale yet. Health and agricultural applications are seen as near term and some have already arrived, sufficient to be persuasive that this will occur. In the longer term, industrial production of seed-stock chemicals is foreseen, but much larger enterprises than this have yet to be tested in any very serious way. And the use of fermentation techniques to replace current techniques in industry is certainly also for the future to provide solutions and answers.

Meanwhile, we now have some spinoff issues that we're not seeing as being quite as dramatic as in fact they are. One of these is the implications for and control over genetic information, i.e., eugenics in the human or agricultural species; or in terms of creating new, large ecosystems (such as would have to be created if there is any colonization of objects in space). What applicability to these questions can we find in the techniques that we are talking about today? And, need it necessarily lead to some kind of control by human beings of their own genetic systems and genetic future. A second issue is that of the impact on previously discussed university-corporation interactions. I don't think that we've solved the problem. There is need for continued attention to this issue.

Thirdly, there is the matter of the regulatory climate for the fledgling biotechnology industry. Apropos, there will be on April 22nd and 23rd, the first Washington conference on biotechnology which will deal most especially with this question. The conference ostensibly will focus on the developments unfolding in Washington, which will establish the near and long-term framework of federal regulations over biotech products and processes. How that regulation shapes up will have a crucial influence on the industry's growth. The conference will open with a discussion of regulatory needs, for example, respecting risks posed by widespread use of genetically engineered products. Is there a need for unprecedentedly tight regulations to deal with these? Put that back 10 years ago, and except for the fact that it's talking specifically about regulations, it

is essentially asking the same question as then --
would there be need for unprecedentedly firm
regulations of some kind? From this question came
the NIH guidelines and the quasi-regulatory
mechanisms that have since governed the area with
decreasing severity. I think this conference will be
of interest to many of you. I think it will be of
interest nationally. So the nature of risk is still
an open one -- is still under discussion. Included
in the conference will be recommendations coming from
the cabinet working group on biotechnology which was
created about 6 or 8 months ago. They have just
completed their review of federal agency rules, and
presumably will provide recommendations for a pattern
involving the NIH, the Recombinant Advisory
Committee, the FDA, the EPA, Consumer Affairs -- the
whole range of regulatory agencies. The key
questions which the conference will address are how
best to regulate research in contained environments
for testing as deliberate releases, and what is the
role of the Environmental Protection Agency with
respect to any releases intended to modify the
environment.

Also on the horizon are direct applications to
the human species. Genetically engineered products
might be one way to deal with disease as you've heard
already from Dr. Friedmann. Growth hormones and the
use of gene therapy for growth retardation, are other
situations. Genetically engineered somatic cells,
for example, have been transferred into mice. There
is the now well-known experiment where rat growth
hormone genes were put into mice and these mice then
grew in a number of instances, to a very large size,
and produced growth hormones not only in the anterior
pituitary but in other tissues as well. Furthermore,
the growth hormone genes were transmitted to
subsequent generations in at least a few instances.
That meant that genetically engineered germ line
cells had been produced -- inadvertently, for the
main purpose -- but nonetheless demonstrating the
possibilities of intergenerational transmission of
genetic characters introduced in mice.) And at the
present time, there are expectations that gene
transfers in somatic cells will deal with the types
of disease to which Dr. Friedmann earlier referred.
These are now regarded as real enough to merit a
special mechanism for review of the Recombinant
Advisory Committee. We note that in the January
Federal Register, there was a relevant preliminary
document published for public comment entitled

"Points to Consider,"(50) analogous to a similarly titled FDA document respecting pharmaceuticals. This was aimed not to control at this time in terms of what is known, but to inform applicants who wish to carry out clinical experimentation in this area, what factors should be considered in making their applications. Now that certainly has some effects in terms of the thinking of applicants and so on, but it doesn't constitute guidelines, even compared to those of the NIH which operated from 1976 until now. The working group is of the view that it is too soon to say exactly what standards are to be applied, but that there should be interaction between it, the public, the Recombinant Advisory Committee, and the community of investigators in finding the appropriate constraints for these pioneer-type, gene therapy experiments.

To summarize, I would say, judging from the past and what we can say of things today, the future of recombinant DNA, of biotechnology in the broader sense, still promises both opportunity and certain severe problems. To move forward prudently will require a combination of initiative (we heard a lot about this today) and caution (of which we've heard little today). Advancing technology will have to be matched by advancing public policy, and innovations in both areas will be required.

Chapter Nine
THE FUTURE OF THE BIOTECH INDUSTRY
Harry Casari

I am not sure if a businessman's view is appropriate because I am not sure biotechnology is a business yet. But a brief look at the industry a will give you a flavor of what I am talking about. There are approximately a thousand organizations, involved in biotechnology. That includes a few big ones, like Atlantic Richfield, here. There are some universities as well in this group. But if you factor these out we are left with perhaps half of these organizations which could be deemed as start - up companies. I address my remarks to these firms and their problems. 200 of these companies are located in California, and some 25 of these are in San Diego. These young, start-up companies have been funded mostly with venture capital. The industry (to stretch the meaning of the word, "industry") is really sort of an anomaly. If you look at the Venture Capital Journal list of 100 venture capital companies which it tracks as a stock index, there are only six biotechnology companies in the index, and of the six, four have deficits. However, Genetech has a P/E ratio of 201 to 1, Hybertech has a P/E ratio of 413 to 1. It's difficult to come up with some remarks that make sense with this mixture. In any event, I think the industry will develop along the same lines as have most other industries.

Most scholars of business history would probably agree that businesses go through the same kinds of developmental stages as does man. There is infancy, old age, and death. The biotech industry is in its infancy. I'm not sure, however, whether it's actually at the fetus stage or at the young infant stage. But it is down there because no one is making any money yet, despite an enormous amount of money being pumped into it. So the problems that you typically will find are right "off the top" kinds of problems, and I will discuss these specifically. These are different incidently, from the kinds of "old age" problems that you had with the failures of the Swifts and Penn Centrals and Packards. I will be discussing the near-term problems of small businesses. There is another factor, also, that's quite significant today -- foreign competition in biotech, particularly in Japan and Europe. This makes it difficult to predict exactly what is going to happen, because the world economy is simply not under our control. Because it is under the control of others, it is even more dangerous to predict our industry's future. We talked about $1.50 gasoline. Now we're down to a dollar gasoline. Who knows where it's going to go next? Some say 75 cents, some say a dollar and a quarter. Regardless, I will speculate concerning events over the next two to four years, and I will discuss only three or four items which I feel will be the emerging key business problems.

Biotech management for one, will require more money and time. Of those 500 biotech start-up companies, only six are public. That means 494 are going to go public as soon as the market gets hot; and what the industry needs is an Apple computer of biotechnology -- somebody to "cure cancer," something fantastic which will make a lot of money -- then everyone will be off to the races. The industry right now is like Del Mar with a starting gate 494 companies long, and everybody shoving horses into these gates. The venture capitalists are up in the stands looking at their tickets wondering which one is going to win. So far, the gate hasn't opened because the market has been dead. They are still clearing off the bodies from last year's track. But as soon as this is done -- bang, the gate's going to open and away the horses will go. That's going to happen, as everything goes in cycles. But before all this happens, there will be a fair amount of litigation. Accountants do a lot of work as expert witnesses and we also get sued frequently. We are the ones who are harassed about money. When somebody goes under, you are not going to kick a dead horse -- you will sue the auditors, who are left with funds. There are many of these suits (and it is fun if you don't have to pay the insurance premiums). I have often served as an expert witness, and worked both for the defense and the plaintiff, so I have a lot of experience in this regard. I think these actions will arise in the biotech industry for two reasons. Some of the young companies are not going to achieve the high promise that was set out for them, and they will fail. In the weeding-out process, a whole segment of the legal profession will be specializing in class action suits against those companies which do not accomplish great things.

The second cause for many of the lawsuits will be the R & D partnerships. These will not come to fruition. I think the reasons they will produce lawsuits are plain. They are technical in nature, and a jury will never understand them. The plaintiff's attorney will have a field day while the defense tries to explain why such and such a thing didn't work out, and why they put so much money into it. Or perhaps why the company itself is doing well, but that this one R & D partnership is not doing as well as others are. Someone will sue and say how come our R & D partnership did not develop this great drug, and the other one did. Certainly we learned something over there, or learned not to go a particular way. "You spent my five million bucks, and learned that that wouldn't work. So pay up." So, I think that there are many problems from a financial manager's point of view. When you're sued and the auditors come in, the attorneys won't tell you whether you are going to win or lose. They never do. We had that fight several years ago. They had more muscle than we did, and we agreed that they wouldn't tell us. When we agreed to that, we "qualified" our "opinion" on a financial statement. And this "qualified opinion," means that it is going to be very difficult to get bank financing, or equity. If you can get it, the price is going to go way up. Also, these suits are going to persist for years. We have that right now in the computer industry, with Osborne, and Eagle, and the Pizza Time Theaters. Every industry goes through it, and biotech will not be any different.

The biotech industry has had a lot of money poured into it and there have been many promises made. At some point the people who have put the money up will get tired of waiting. I hear today of waiting eight to ten years for these things to pay off. But how many venture capitalists are going to sit around on their hands and wait that long? They usually oust the managers after a couple of years if things are not on track. Although I'm being somewhat hard and a little facetious, there truly will be great pressure brought on the industry to show results because people who put up the money must show results to their investors. They must keep raising money to put into new companies, and their track record is only as good as their last venture. You can see that pressure already.

There is a company in this industry, which issued an annual report a few years ago. The report read as though it was an operating company. There was an earnings per share increase in revenues; there were also the graphic and tabular presentations and the president's letter. In fact, they did have earnings that year, and they made money. From all this, you could readily believe that the company was in a permanently operating cycle with dividends and increases in stock, etc. It was great. It was well done and it was truthful. The next year, the same company, the same annual report, but suddenly there were no numbers. Where were the numbers? They are on page 27, with a description of ivory tower research ("We will save the world" stuff) and associated up-front expenses. When you finally get to the numbers, you learn that the company had a horrible loss attributable to this R & D. So in one year it was an operating company making money and in the next it is back into the R & D mode and it was losing money. The question was, what was the company? Was it a safe investment vehicle for conservative investors, or was it a risky R & D venture? I don't think management focused on that question. I think they got swept away in the first year -- they started to make money ("look how happy the investors are going to be "), and poured it on. So then in the second year, they decided to do some more research ("we're going into this area") and boom, there went the earnings. Somebody investing in year one might sue when he received year two's report. And he will sue if year three does not look pretty good.

So I think this illustrates that management will have to stand up and call it as it is -- it must differentiate between scientific and economic accomplishments. The industry is peopled by those with academic backgrounds whose natural bent is research. This is the basis for the industry. And I think, that being the case, it is easy for them to get carried away with their scientific accomplishments. But that must translate into a bottom line if you are to tell the world about it. And unless that line is drawn carefully, a shareholder's lawsuit can result. Recently, Bill Lerach, of Milberg, Weiss, Bershad and Specthrie, one of the premier class action law firms in the United States told me about the Wickes case. One of the things that Wickes(51) had in its annual report was a statement that their lumber division was "number one." (And this is an example to show you how easily you can get in to trouble) So for the next three years, the guys down in the art department figured that Wickes was indeed "number one" and insured that those annual reports bore this same description. What management meant was that their revenues were number one. They were bigger than any other company of that type in the industry. Unfortunately, that company had a large net loss for each of those years, which they testified to on the stand. The jury thought that that wasn't too fair, i.e. that it should have meant "number one" in net income also. Naturally, Wickes lost the case.

So, although most biotech companies are now private, as public offerings become common, the element of careful management will become essential. Also, as these companies become increasingly commercialized, they will need more people. Different kinds of people in different areas, e.g., financial people, marketing people, and operations people. They will have to compete with other industries for these people.

Money is now in short supply for many biotech companies. Innovative compensation plans will have to be developed to attract these people. Opportunity capital may be used to pay them, and the promise of fortunes to be made can lure them out of complacent positions with the big drug companies, etc. I will merely suggest here some types of plans that I think might work today (though the IRS may say "no" tomorrow). Some programs allow an employee to purchase a stock at market value, less a prescribed amount, and then to have several ways to benefit when he sells it. He thus buys at a very low rate. Another possibility is convertible debentures. The employee buys these from the company, which lends her the money for these, and which debentures are then convertible into the stock of the company. It's a low risk way to get the employee into something. There are also transferable warrants, where you sell employees warrants to buy stock, but they never buy the stock. Instead, when the stock goes up, they may sell the warrants back to you. The advantage here , too, is that they get in for a very low investment. Another is the creation of a subsidiary. Perhaps you want to take on a researcher. He's got a perfect idea, and he wants a lot of money for it, but you do not have a lot of money. So you say, "we'll set up a little subsidiary, you go up there and do your thing, and we'll sell you 20% of this subsidiary. Now when the thing works, we'll merge the subsidiary with the parent and you exchange the stock in a tax-free transaction, and then sell our stock at a capital-gains rate."

There are a host of tax and financial aspects, to any of these plans. But the industry must get heavily into these or similar plans if it is to succeed. When Apple needed someone, they could go into Pepsi, and pull Scully right out of Pepsi and offer him a million dollars a year and a number of other rewards. But that was a young teen-age kind of company that was generating a lot of cash. This is an industry where there are toddlers, they are walking, grabbing onto things, and they don't have that kind of cash to throw around. The President of Pepsi-Cola will have to come and sell whatever you guys are going to develop --if you want to sell.

This brings us to the third problem facing the industry, the need for professional, and timely marketing. There is an article in Computer System News Business and Finance which bemoans the fact that this year the stock market for high-tech stocks went to pot and they're all down and saying, "God, it's a handwringer -- why did this happen to us?" The writer says, "the bubble burst when the technology industry became a commodity industry. Commodity industries are generally unattractive to investors who buy when the commodity is in short supply, not when there's an over supply." He was talking about computers. Investors thought computers were hi-tech and suddenly computers were being sold by Sears Roebuck. The article continued, "the real mistake is that the things we valued highly were not really technology. We confused real technology with making a consumer product. There was not enough differentiation between proprietary technology and something that has to slug it out in the marketplace.

We all underestimated the importance of marketing and advertising to the success of these companies." I think that is point three -- the problem is marketing. And I think this is a problem of every young start-up company.

You have got to bring on real professionals. Ken Tingy of Ventana Growth Fund, a San Diego Venture capital company, told me that they were thinking of making an investment in a biotech firm, and had spent two weeks investigating it. I said to him, what is the biggest problem you've found; in general, what's the biggest problem? In response to my question he stated that it was marketing, or a lack of marketing focus. Let us assume an arbitrary division of the biotech market, divide it up into agriculture, pharmaceuticals, and human applications. I think what happens is that there is a tendency to attack all these areas, or even to attack only one of these areas, like agriculture without breaking it down into plants and animals. I think that you should really be breaking agriculture into animals, swine, maybe baby pigs, etc. You really need a rifle when you are marketing; you can't use a shotgun. I think that this lack of focus in your own companies is pretty common.

Thus, there is the need for marketing discipline in your industry. This is to bridge the gap between the basic research and the envisioned applications. So great, in fact, may some of the technological leaps be, that it is more challenging sometimes, to define the limitations to successful applications. ¶A great leap is made that if this works, it will do all these things, so let us go try to sell it to all these people.' You have to have discipline - that is where you need a professional marketer who knows how to analyze a market. She's able to say "here's what we are going to go after, because this will make a lot of money for us." Marketing cannot be relegated to the weak-sister role.

I think another problem here is that a lot of the people in the industry, being from academia, probably do not appreciate that marketing really is an art. It is as demanding, and maybe even more so that their R&D. You really have to know what you're doing to market successfully.

A related problem may be that there is a hesitancy to get away from the biotech-as-science image. Being thought of as a biotech company has a lot of stature, it has a lot of acceptance because it goes with intellect and academia and a new way of discovering things to cure mankind. Just think of Jonas Salk's reputation in the community. Who woouldn't like to have that? I'm an accountant, and I would even trade my reputation for Jonas Salk's. So, when you are starting out, and you're a potential Jonas Salk, it's hard to say when you go home at night, when asked what did you do today? "Well, you know, I invented this stuff and if you put it in a hog's rear, it's going to make a lot more ham." The response is not going to be a Nobel prize, it's going to be: "Go wash your hands before dinner." That is a real problem. I used to audit a lot of companies who were in that business, the animal business, animal ills. It was a smelly, obscure, but <u>extremely</u> profitable business. However, the guys were never very popular at the country club.

My final point is sort of a twist on marketing. I frankly think that biotech has done an amazing job of selling itself. It has convinced a tough American audience that there is great promise for this industry. It has the whole venture capital community eating rtight out of its hand. (If you do not believe it, just look at how much money it has raissed already).

But the companies which have raised this money, now need to use the same talent in the marketplace, because the products of biotech must also be successfully sold. The ultimate value of anything is whether or not it provides a benefit to society. In a capitalistic system, this test is met in the marketplace. I recognize, that there are exceptions to this statement, e.g., the treatment of some diseases will never be profitable making them -- we will have to go after that on a humanitarian basis. But I think these five hundred or so companies were funded with the idea that they would ultimately be successful in the marketplace.

Chapter Ten
THE FUTURE REGULATION OF BIOTECHNOLOGY
Robert A. Bohrer

An issue that has been raised here a number of times today is the central function of the lawyer in the development of biotechnology. A common conception is that the lawyer in high-tech or biotech is primarily concerned with proprietary rights. That is a relatively benign image of the lawyer's role. An alternative conception, and a much less flattering one, is the image of the faceless bureaucrat who is impeding the economy, impeding your work, your profits, and in general mucking things up. We have been in an anti-regulatory climate for some time, during which there has been a big push for deregulation, and it has been very popular to view the lawyer-regulator as the devil. What I hope to do is to explain why there is regulation, why there will continue to be regulation, and even try to convince you that, at least in some areas of biotechnology, what is needed is even more stringent regulation.

There was an article in this morning's Los Angeles Times (2/28/85), reporting that the D.C. Court of Appeals ruled that the N.I.H. must take a harder look at its approval of the ice-minus bacteria for small-scale field testing.(52) The court did not require a full-scale environmental impact statement, but it did direct the NIH to seriously consider whether an environmental impact statement was necessary, and what the environmental considerations might be. This is an obvious example of lawyers and regulators applying the brakes to biotechnology.

What I will undertake is to explain why those brakes are there, what the nature of those brakes is, and to compare the way the brakes are being applied to biotechnology with the general theory of risk regulation and in comparison with the regulation of other technologies.

Perhaps the single most important factor in the regulation of biotechnology is one that has been referred to again and again here today -- that is, the paucity of data. Dr. Pinon spoke about the absence of data in the early days leading up to the Asilomar conference, which led to the conference's consensus ban on experimentation of particular types. Similarly, the first NIH RAC guidelines were very stringent, and it was the paucity of data that led to that stringency. Yet what happened is that, despite the stringency of those early guidelines, data began to accumulate and experience grew, which led to relaxation of the stringency of control. Each set of NIH guidelines has allowed researchers greater leeway than the one before.(53) I think that this process is an important part of what may be referred to as the life-cycle of the regulation of new technology. In the beginning, faced with an absence of data concerning a variety of possible risks, the inevitable response is a very stringent regulatory framework. As data accumulate, even with that stringent framework, experience enables a better assessment of the actual magnitude of the risk presented, and the regulatory approach is reassessed and loosened when appropriate.

This general life-cycle of technology raises several very important issues. The first question is how can the risks of new technologies be assessed before they can be quantified with any degree of confidence? The second question is how can we translate these unquantified risk assessments to an appropriate level of regulatory stringency? A third question is how is the general concept of regulatory stringency expressed in specific legislative and administrative choices? Finally, how does the regulatory framework adjust to experience and become progressively less restrictive? I will suggest a model of regulation that may provide answers to these questions.

We have already experienced a significant enough portion of the life-cycle of recombinant-DNA technology to see the pattern established. The FDA, which in general has a very stringent regulatory mechanism, has managed to collect enough experience that it is already beginning to loosen its control of recombinant-DNA products. We know, therefore, that the process does occur, but not necessarily why or how.

The first question, as I have already suggested, is how are risks assessed before their magnitude can be ascertained with any degree of confidence? Risk is most commonly defined as a product of the probability of adverse consequences and the value of those adverse consequences. More simply, it is the answer to two questions -- how likely is it that there will be bad results, and how bad are those results? This is a very traditional approach to risk analysis. Under this approach, a ten percent chance of a $100,000 loss is the same magnitude of risk as a fifty percent chance of a $20,000 loss. It is this traditional concept of risk that raises the basic questions of how risk can be evaluated in the absence of any meaningful quantification. I believe that there is a way of talking about risks in a meaningful way before you can calculate the risk with confidence. To do that, you must accept the assumption that risks have attributes other than magnitude, and that those other attributes can reasonably determine how we approach those risks. We might call this qualitative, rather than quantitative, risk assessment. All new technologies must undergo such qualitative assessments at the beginning of their life-cycles.

Although we have begun to acquire a degree of confidence about the risks of some kinds of biotechnology applications, we have much less confidence about others. For those, we must make an attempt at qualitative risk assessment and proceed to create an appropriate regulatory structure. As Dr. Grobstein said, and others have also acknowledged, the biohazards of experimentation in a contained laboratory setting are well known, well understood, and, apparently, are being dealt with appropriately. The technology that is less well understood is the technology of gene therapy which Dr. Friedmann discussed, and I believe that even further down the scale of confidence in quantitative assessment is the technology that requires environmental releases of engineered microbes, for agricultural and other functions, such as Chakrabarty's oil eaters.(54)

How should these less well understood risks be characterized? There are five factors which I believe determine qualitatively the way in which risks are regulated. These are the market nature of the risk, the complexity of the mechanism by which injuries occur, the perceived patterns of injury, the new/old factor, and what I will refer to as the political x factor.(55) I will briefly describe each of these qualitative risk factors and their implications for the probable stringency with which representative biotechnology applications will be regulated. I will then discuss briefly the relationship between the general concept of regulatory stringency I will be using and particular legislative and administrative choices. I will conclude by using that dual model of qualitative risk and regulatory stringency to make some general predictions about future regulatory directions.

The first is the market nature of the risk. That is, some risks seem to be primarily a result of voluntary transactions between two persons, and are born primarily or solely by the parties to those transactions. These are market risks, and however imperfect the market may be, they are treated differently than non-market risks, or those risks which are primarily directed at third parties, without their consent and perhaps without their knowledge. I believe that biotechnology presents both kinds of risks, and that as with other technologies, we are much more stringent about the regulations of non-market risks to third parties. For example, fermentation technologies, in which genetically engineered microbes are used to produce such products as insulin or human growth hormone, create risks which are primarily market risks between two parties. The FDA's role there is to correct the structural imperfection of the market that arises from the usually large disparity in knowledge between the two parties -- the producer who can run sufficient tests to determine the safety of his product and the consumer who cannot.

Gene therapy also seems to be a risk between two parties. I say "seems to be" because the Points to Consider(56) that the FDA recently released raise the issue of the extent to which gene therapy poses risks to third parties, presumably because of the use of living, and therefore possibly infectious, vectors. (57) The FDA is responding quite appropriately in singling out third-party risks for special attention. Although we do not presently have reliable data on the risks of gene therapy to the patient or to third parties, it is reasonable to be more concerned about third party risks, and to subject those risks to a more stringent regulatory framework.

Environmental releases, or agricultural applications present predominantly third-party risks. Sure, there is some risk to the farmer whose strawberries are being sprayed, but the nature of such applications, as the recent regulatory pronouncements made by the interagency task force clearly emphasize, is that they are not contained. If you spray a crop with a pesticide which is an inert chemical, and it's a bad pesticide, the crop owner will suffer, perhaps the applicator will suffer, maybe the people who produce it will suffer; it is even possible that residents of the neighborhood will suffer, but even that is a relatively contained population that is exposed. However, when you spray a crop with an animate, or living pesticide or plant regulator, it's not just his crop and the immediate neighborhood that is at risk; it is, in fact, all of us. It is not necessary to be Jeremy Rifkind or a member of the flat-earth society to accept this statement.(58) I do not necessarily mean by saying that we are all at risk that such experiments threaten us with catastrophe or the end of life as we know it, merely that whatever risks such experiments pose cannot be contained. It is the nature of these organisms to reproduce; it would not be worth spraying them on crops if they were not hardy enough to survive in the environment. The potential is the creation of another gypsy moth. A useful organism in the wrong ecological niche can cause damage over very large areas. The risks are inevitably to third persons as much as to the parties to the transaction. Thus, for qualitative risk factor number one, market nature, we can identify environmental releases as most troublesome.

Factor two is the complexity of the mechanism by which injuries may be produced. This is a difficult factor to elucidate, but one which I feel has a definite affect on our response to risk and to regulatory stringency. It can be related back to the issue of the market, because complexity translates into information costs and transactions costs. The more complex the ways in which injury can occur, the more unforeseeable are the kinds of consequences, and the less confident we are about our ability to predict the magnitude of the risk; therefore, a more stringent regulatory framework is required while information is gathered. Again, we can illustrate with biotechnologies. Fermentation technology seems to pose relatively straightforward risk. The risk of impurities or toxicity is direct, and can, apparently, be dealt with by controls and tests that are already available. The complexity of the system is much the same as for other methods of pharmaceutical production, and that seems to be the approach the FDA is taking.

When we move to an examination of gene therapy, I think Dr. Friedmann would be the first to acknowledge that we are dealing with a much more complex system. The retroviral vector is a complex one; the way in which the vector invades the cell is complex; the integration of the genetic material and the production of the new protein are all complex processes, each of which can produce a bad result. I would not pretend to be able to list them, but it seems unarguable that the potential sources of injury are far more complex than with fermentation processes. That would seem also to justify the position of the FDA that it will not treat gene therapy like a new antibiotic, that approval of gene therapy will require a proof of efficacy and safety beyond that which would be required for a new antibiotic. I think that in part this should be attributed to complexity of mechanism.

Finally, there are environmental releases. Again, wholly apart from magnitude, the complexity of the mechanism by which injury might be produced is even greater than for gene therapy. This might seem to be a rather difficult assertion to support. After all, you might respond, how can one compare the complexity of modified E. coli, which have been studied to death, with the complexity of retroviruses? The answer is that the complexity of the injury-causing mechanism is only slightly related to the complexity of the organisms involved, and is more closely related to the complexity of the system into which those organisms are introduced. While the mechanism for gene therapy can produce injuries in that extremely complex system known as the human body, I think it is fair to say that we now have a better understanding of that complex system and of the way inadvertent adverse consequences can occur than we do of the ecosphere, in which environmental releases can cause harm. The ways in which things can go awry in the ecosphere are far more difficult for us to understand, and therefore, what I mean by complexity is far greater than even in gene therapy. I think the classic proof of this contention is our experience with DDT. Although DDT had been tested in the laboratory repeatedly and found "safe," the first indication of trouble was the dwindling bird population due to DDT-caused thinning of eggshells. That was indeed a circuitous and unpredictable harm, which revealed the fragile ecological complexity of our environment.(59) Thus, with respect to the second qualitative dimension of risk, complexity of injury mechanism, the result is again to highlight environmental releases as applications which seem to require the most stringent regulatory scheme.

Factor three is what I will call the perceived patterns of injury. This factor is a rather simple one -- do we see the pattern of injuries which are likely to result from the technology as sporadic and episodic occurrences, or do we see the pattern of injury as being concentrated in more infrequent mishaps of considerable magnitude?(60) We are far more willing to tolerate 50,000 deaths per year from automobiles, which kill us in isolated events, than we would be willing to tolerate the possibility of one or two nuclear power incidents per year that would result in even one-half that number of casualties. To be sure, this quality of risk factor does not operate independently of the others, but it does have an additional impact of its own.

Here again, we can see some clear implications for different biotechnology applications. At the time of the Asilomar conference, the prevailing perception was very clearly one of the big bang -- the Andromeda strain that would bring a plague that could destroy mankind. The passion expressed at Asilomar certainly did not stem from the possibility that there would be a significant increase in the mortality rates for the scientists and laboratory technicians who were exposed to the chemicals and products of genetic engineering. We are clearly far more motivated by the potential catastrophe than continuing, intermittent death. Dr. Grobstein has said that there have been no injuries due to genetic engineering research, and I think that is a position that has been accepted by the NIH as a basis for the relaxation of their guidelines. Nevertheless, I believe it is quite possible that there has been an increased incidence, at some low level of magnitude, of some disease related to the exposures tolerated by biotechnology workers. It may or may not be occurring, and it may or may not rise to a level which is epidemiologically detectable or significant, but whatever is the case, it is simply unlikely to arouse much concern or call for regulation unless the numbers become really big.

To conclude my discussion of factor three, it is clear that when the pattern of injury is perceived as focused on single events of large magnitude, even with very low probabilities of occurrence, we are more likely to demand regulatory stringency than for the pattern of injury which is episodic. It is also clear that the perceived injury patterns for the various biotechnology applications are very different. There is perceived to be little or no catastrophe potential in fermentation technology or in gene therapy, but again, environmental releases raise a far greater issue as to the pattern of potential injury. Even though the chance of injury may be very, very small, the injury might have a more global effect than a sporadic one. Again, to draw on the analogy to DDT, we have learned that we are not very good at making environmental predictions, and that persistent substances can have wide-ranging unforeseen consequences. The result of this third factor is again to point to environmental releases as the application of biotechnology which calls for the most stringent regulatory regime.

The fourth factor has been discussed by others as the new/old factor.(61) We are far more stringent in controlling the risks of new technologies than we are for old technologies, even where the risks are otherwise the same, both qualitatively and quantitatively. A clear illustration is the difference in the way we treat risks generated by coal-burning power plants and the way we treat the risks of nuclear power plants. I believe that the reason for this differing treatment is that it is far more difficult to trade risk against uncertain economic benefits than against established economic rewards. Here too, the implications for biotechnology are clear. Much of the discussion today has forcefully underscored the uncertainty of the economic rewards of biotechnology and the difficulties that will be faced in obtaining economic benefits. We hold out the great promise of curing all of mankind's ills and making this a better world, but that day has not yet come -- the economic benefits have not yet started. The result is that we still have an industry that is unborn, not yet even an infant, and therefore likely to be treated stringently. We will accept much less risk from biotechnology than we do from the petrochemical industry, where there are also tremendous risks to human health, but where the existing economic returns are a powerful force.

The fifth factor in my model of qualitative risk analysis is a somewhat different one. The first four factors are attributes of the technology that impact on the way its risks are perceived and treated. The fifth factor is not truly an attribute of the technology but a variable attribute of ourselves, which I call the political X factor. The stringency of our regulatory framework will be affected by where our society is in the cycle of ups and downs, well-being and need, optimism and fear. Just as the Dow-Jones Industrial average rises and falls, so do our feeling about regulation. It certainly is connected to our sense of well-being, and conversely, our general willingness to take risk in exchange for material benefit. We are far more willing to regulate stringently when we are feeling well-off than when we are not. Thus I think that we can perceive a cycle in our attitudes towards regulation, and that we are down quite a bit from the high-water mark of the early 70s in our tolerance of regulation. However, I think we are also moving back up a bit from where we were in 1980, when deregulatory promises, among other things, swept a president into office. While it is difficult to reach firm conclusions about the implications of our general social consensus about regulation for biotechnology, I think it is fair to say that we will be somewhat less stringent in controlling biotechnology than we would have been had these issues arisen a decade ago.

In summary, I think that these five factors do affect the way in which risks can be characterized qualitatively, even before their quantitative magnitudes can be determined with any confidence. Each of the three biotechnology applications I have been discussing varies a great deal in the qualitative nature of its risks. Fermentation processes present risks primarily through the market, not to third parties. The mechanism by which injuries may occur is not terribly complex. The perceived pattern of injury is likely to be episodic. In short, we are unlikely to find any demand for new, more stringent regulatory controls for fermentation.

Gene therapy is somewhat different -- chiefly in the complexity of the injury-causing mechanism and the somewhat less settled question of risk to third parties. The result is that we are seeing, and will continue to see, a fairly strict regulatory scrutiny, stricter than for other medical therapeutic applications. That will continue until the qualitative dimensions of the risk are changed by quantitative data as to the actual magnitude of the risks involved.

Finally, there are the proposed releases of genetically engineered microorganisms. Here, I repeat the conclusion that the risk is primarily to third parties; it is very complex; the perceived pattern of injury is global rather than episodic; and, therefore, we are likely to see the most stringent regulatory framework and the strictest scrutiny applied to these applications. Again, this will continue until we have sufficient experience to conclude that we know what the actual risks are, how to control them, and how to proceed with the development of the technology.

Up to this point, I have been describing the qualitative attributes of risks which, in the absence of quantitative information, should lead to a more or less stringent regulatory framework. Now I shall briefly discuss the way in which regulatory stringency, as I have been using that phrase, translates into particular Congressional and administrative choices. Congress plays the most important role in this process, when it allocates the burden of proof, when it defines the regulatory standards, and when it sets the standard for judicial review. Let me briefly discuss each of these three factors.

The burden of proof is probably the most important choice of all. While lawyers generally view the allocation of the burden of proof as very important, it is far more important here because of the underlying principle which I have been discussing -- the absence of data and the existence of uncertainty.(62) When you have to prove or disprove the uncertain, the burden of proof is fatal. If you require the agency to prove danger, in the absence of data, then the agency will be unable to regulate. Similarly, if you require the outside party to prove that its product of activity is safe, when data is unavailable, then the burden of proof is inevitably fatal. Obviously, what I am calling a stringent regulatory framework is one in which the burden of proof is on the outside party seeking approval.

The regulatory standard can be set either at a level which seeks to eliminate risk, and thus phrased in terms such as "will not endanger" or "will be safe," or can be set at a level which is open, to a greater or lesser degree, to the balancing of costs and benefits, i.e., "will not unreasonably impair" or "to the extent feasible." The FDA standard has been one which ostensibly eliminates risk by requiring a finding of safety, the EPA standard under FIFRA has been a somewhat less risk averse standard of "unreasonable risk."(63) What I would term a stringent regulatory framework is one which attempts to minimize risk, is relatively insensitive to costs, and thus uses a standard such as "will not endanger."

Finally, we come to the standard of judicial review. Here we have the point at which Congress's choice of regulatory stringency intersects both the agency's own use of its discretion as well as with the court's view of Congress's intent. To some degree, Congress can shield an agency from reversal by the courts, if it provides a minimal "arbitrary and capricious" standard of review, instead of a more rigorous "substantial evidence" standard of review. Congress's choice of a minimal standard of review, coupled with an agency's desire to exercise its discretion with maximum rigor, leads to the most stringent regulation. Congress's choice of the substantial evidence standard weakens the agency's ability to withstand judicial review and gives the courts a much greater role in the administrative process. However, the minimal standard of review, in terms of regulatory stringency, is a dual-edged sword. It allows an agency to get tough, but it also allows the agency to back off. The NRC is a good example of an agency which, in the eyes of many critics, has used its discretion to back away from stringent regulation. Nevertheless, I would believe that an appropriately stringent regulatory standard for the regulation of environmental releases of genetically engineered microbes would be based on the arbitrary and capricious standard of review, coupled with an appropriate attitude of determination on the part of the agency involved. The substantial evidence standard is likely to be a formidable obstacle to stringent regulation for the same reason that the burden of proof would be fatal -- regulation requiring substantial evidence to be sustained would be impossible until substantial evidence is available.

What I have just described may be somewhat dismaying to an audience of business entrepeneurs and scientists, suggesting a regulatory framework in which nothing could be done because nothing can be proven safe and nothing can be proven safe because nothing can be done. While this is a catch-22, it is not a permanent one. The experience of the NIH in revising the RAC guidelines demonstrates the ability of agencies to adapt. While that took place in the presence of a constrictive, self-imposed, nonstatutory framework, that difference is not fatal. To recall the example of the NRC which I just referred to, the ultimate value of the administrative discretion inherent in the arbitrary and capricious standard is the ability of the agency to adapt over time. Nuclear power was subjected to the most stringent regulatory framework. The burden of proof is on the applicant for the license, the standard is "safe" and the agency is shielded from judicial second-guessing. Nevertheless, the agency has loosened its grip on the industry as it gained confidence in the underlying technology. Some would indeed argue that the pendulum has swung too far and that the agency is abdicating its responsibility. What is clear is that stringent regulatory framework is not a death-knell for a new technology, but that its development depends on the agency's accumulation of experience and its exercise of discretion.

I have now set out the basic features of my model of qualitative risk analysis and regulatory stringency. What remains is to ask whether the emerging regulatory framework for biotechnology fits the risks as I have described them. The regulatory framework is rapidly becoming clear. The FDA will have jurisdiction over fermentation products in the form of pharmaceuticals, biologics, diagnostics, all of the medical applications which have been discussed. To the extent that the technology is new, they are taking a harder look, and to the extent that the technology becomes well established, they tend to loosen up a bit. I do not foresee any particular change in that pattern. The FDA has taken control, with an adequately stringent regulatory framework, and it is exercising its discretion in proportion to its experience with the technology.

The FDA also has jurisdiction over gene therapy. In the NIH Points to Consider, it is clear that gene therapy will be treated more stringently than other new drug applications. For example, it suggests the use of primate testing, which is not a standard IND requirement. It also expresses concern about the risks to persons other than patients. What is important is that the FDA's regulatory framework is a very stringent one, which gives the agency adequate power to deal with these risks.

Finally, we have the agricultural and industrial applications of biotechnology which entail the release of genetically engineered microorganisms into the environment. This application is obviously the one which is central to the concerns of my model. It is the area where the risk is qualitatively the largest and where the regulatory control should be most stringent. How will the EPA regulate it? It has jurisdiction under two separate statutes -- FIFRA for the agricultural applications and TSCA for the non-agricultural applications. FIFRA is a statute that provides for relatively stringent regulatory control of new substances, and the preliminary approach of the EPA indicates that they intend to exercise their discretion relatively rigorously. For example, ordinarily, small-scale field tests involving less than ten acres do not require prior EPA approval, but the EPA has announced that it will require submissions of data prior to small-scale field tests of genetically engineered microbes. Although the regulatory standard might be a little tougher, FIFRA is probably an adequate regulatory framework, if the EPA wishes to use it fully.

TSCA presents a different story. It is the general framework for the regulation of the manufacture of all new chemical substances, and thus would apply to non-agricultural uses, such as Chakrabarty's oil eaters, if genetically engineered microbes can properly be classed as new chemical substances. The EPA has announced that it believes that it does have such jurisdiction, and that, as with FIFRA, they are going to take a hard look at applications for the approval of new microbes. The EPA's TSCA regulations have exempted R&D testing from the agency's prior approval requirements, but the agency will not extend such automatic clearances to environmental releases of new microbes.(64) So they are using their discretion, within that framework, to regulate stringently. However, the framework itself is a far less stringent one than FIFRA. Applicants for product approval need submit only limited data,(65) and the burden is thus on the agency to produce reasons why the product is dangerous and to support that conclusion with substantial evidence. When push comes to shove, the EPA will have a hard time justifying a decision to reject an application for R&D field testing of these products, if the applicant wants to challenge the agency and has not generated the data that the agency would like. In short, the statutory framework is inadequate for the task. In reviewing the current regulatory framework, it is TSCA which is most obviously inadequate and most clearly demands strengthening to deal with new developments in biotechnology.

In conclusion, the real center of the future debate over the regulation of biotechnology will be over the control of environmental releases of R-DNA products. That will be a continuing controversy, with calls for a more stringent framework continuing to be heard, and, at some point in the future, an alteration of the existing framework.

COMMENT: FUTURE TRENDS
Robert A. Bohrer

Let me take again the privilege of summing up. We started with exchanging our basic overviews of what scientists do, what business people do, what lawyers do. I said then that my perception of what we do is that we solve problems; that we solve different kinds of problems and that those problems are all involved in the process by which technology goes from pure research to world application. It is a mutual enterprise - a joint enterprise, in which we are all necessary parties. As we moved into current problems, we began to look at the way our problem-solving approaches operate in those areas that are currently our most uncertain and therefore our greatest concern: gene therapy, the business-university relationship, proprietary issues, regulatory issues, and so on.

In this last session, what we have done is reverted to our most basic problem-solving mode, to create models that are drawn on our experience as problem-solvers within our own discipline, to say a few words about what is to come. I, obviously, have one model for problem-solving, Harry Casari has another model for problem-solving and therefore another model for what problems will arise, and Cliff Grobstein has yet a third. I learned from each of those steps in this interchange. I learned something about other people's models and the kinds of problems they predict and the kinds of variables they consider.

QUESTIONS AND ANSWERS

Question: While it may be implied in your analysis, there seems to be not enough attention paid to benefits. How can you talk about risks and a stringent regulatory regime, without weighing those risks against the benefits. An example might be that if you look at the world as a whole, and not the United States, an increase in agricultural production might do far more good than anything else we can think of, and yet that is where you would place the greatest restriction on improvements that, by nature, must be released to the environment.

Answer: Bohrer: I think that question was directed at me. The question is to what extent am I giving short shrift to benefits, especially using agriculture as an example, because it is there that the potential benefits are so enormous. I don't give short shrift to benefits. I think what my model suggests is that benefits can only be evaluated in a meaningful analysis when there is also meaningful data about the risks. And so there is the need to develop the experience curve in dealing with these new technologies. At the beginning, the very reason we treat these technologies stringently, is because we can't weigh uncertain benefits and uncertain risks, so we attempt to characterize those risks. I have, obviously, a more sophisticated model for characterizing risks than I do for characterizing unknown benefits. I don't mean it is because the benefits aren't important, but what I am suggesting is that it is a two-step process. First, you characterize the risk and regulate from that basis, while you gather the information necessary to weigh the risks and benefits in a meaningful way. Then it is possible to decide what you want to do. And while that is a slowing up of the process, that many of you, I know, will lament and bemoan, it is not an end to the process. It will happen a little bit later, but I don't think that caution is something that we want to dispense with, going very slowly until we have a handle on both the actual magnitude of those risks and the actual magnitude of those benefits.

Question: I think that if that question could be taken a bit further, is what is the normative basis for not weighing benefits at the same time? Are you saying that is how we should behave?

Answer: Yes. And I think it is a question of at what point do we address benefits. I think we don't address benefits at the very early stages, and I don't think we should address benefits at the very early stages of creating a framework. Because the benefits are as uncertain, the magnitude of the benefits are as uncertain as the magnitude of the risks. So, until we can talk about the magnitude of the risks, I don't want to talk about the magnitude of the benefits. All I want to talk about is the characteristics of these risks in this non-magnitude-driven way, which is your interim step to learning about the actual magnitude of the risks and benefits and then responding accordingly. You have got a first stage of regulation which is creating a framework that responds to characteristics of the risk, and right, not the characterists of the benefits. And you have a second stage during which knowledge accumulates, and then you can address both the actual magnitude of the risks and the actual magnitude of the benefits, but I don't think you should address benefits until you can say something more meaningful about risks in terms of their actual magnitude. Now, that is my position, and I suppose there is another question. That is my normative position. Now my empirical position is that that is, in fact, the way we work. We do tend, when we are confronted with these kinds of problems, to respond in that way. I think of Asilomar, where it wasn't lawyers, by and large, who were involved, but it was a group of decision-makers faced with a similar problem, and while they could forsee those benefits, the framework they adopted was a very stringent framework, because they didn't get any of the data to assess the magnitude of the risks. Now, Cliff, disagree.

Grobstein: Yes, I think I have to disagree. It seems to me clear that what you have said about the analysis of risk is appropriate. I think, and useful. And I think that your point is that it is important to be able to say something about risk even before you can speak to magnitude. But I think the same thing applies to benefits then. I join with the questioner in expressing concern. That if a regulatory process is driven by risk, then obviously it is going to be far more conservative and stringent than if it is driven by benefits. And the fact is, that historically, it is the benefit that comes first. We are not motivated to generate risks, except in games. We are motivated to generate benefits. And, in fact, the people who came together in Asilomar were all convinced that what had happened and could happen would be a benefit. They just assumed that. They didn't ask a sophisticated question like: can we specify the benefits without having more data than we currently have? It seems to me that your analysis, to be even-handed, has simultaneously to address both questions. The view historically, and maybe that is because you think legally, was to primarily think mnore about risk. But if, in fact, this passes on, or we pass on and leave it, then to the present, there will be a regulatory process, it will be over-driven by this risk analysis. And I think it is fair to say that it should be a compensatory address to the benefits side.

Bohrer: Again, I disagree. I think I want to distinguish the process from the framework. Because, I think, it evolves over time. The framework is, I think, driven by risk. The process adjusts to knowledge about benefits as much as it does to knowledge about risks. I think that is illustrated by what the FDA has done in the area of fermentation technology. The early response was case-by-case cautious, and it is my impression that they are now responding to the magnitude of what they see as the benefits, to expedite things. Vince, please.

Frank: I would disagree and say that what they are responding to is the magnitude of the risk. Theoretically and I think in practice, the FDA, currently is a pure risk analysis. You have to show us your product is safe, the manufacturer must show data, and data minimizes your risk to the extent that you show that nobody gets hurt by the product. So, from my perspective, at least in the FDA regulatory process, benefits are not as central as risk. Another question is how does the current regulatory framework differ from what you propose?

Bohrer: What I am suggesting is that what the current regulatory environment lacks is a meaningful regulatory structure for non-agricultural environmental releases. TSCA is not a meaningful regulatory structure designed for this purpose. The FDA does have a structure designed for other purposes that seemed adequate. The EPA does have a structure under FIFRA for agricultural uses that seems adequate, but TSCA is not what would occur to one when you want to write a statute to deal with this.

The farthest thing from what would occur to Congress to enact for environmental release would be TSCA. To again respond, because I guess I am becoming the focus of a little controversy, as I had hoped (and I will hope that someone will ask a question for somebody else) it is a process that takes place over time. I don't think it's thinking merely legally, but if it is, I'm not going to say, "I have sinned and I will go and sin no more." I will certainly continue to think legally. But I think that the regulatory process is at the beginning risk-determined, and that it changes over time as the magnitude of the risk becomes more knowable to being driven both by risks and benefits, and that is the example again of the NRC, maybe not the FDA. If this is for me, I'll answer it.

Question: I hope I think legally, too. But I think your example of the NRC is not well taken. The regulatory regime became more restrictive over time, not because perceived risks were greater, but, I think, because of perceived benefits was regarded as being lower than originally believed. And that was not a relaxation. It was a definite tightening, and I don't think realistically so, of regulations.

Bohrer: We have a disagreement about what the NRC has done. I think that actually you're right, depending on which point in time and which particular NRC action you look at, and I think there was a loosening of the NRC standards that reached the high water mark, at least with respect to some of their jurisdiction, in the Vermont Yankee case,[61] their decision to determine that were no environmental impacts of low level waste storage. I think their decision to do that by rule was a real loosening of NRC standards. I think they are tightening up again in response to Three Mile Island[62] and other things that bring pressure to bear on them, so I think there is a mixed bag with respect to NRC experience.

Question: A question for Dr. Grobstein. It took science about six years, roughly, to loosen up the regulations, just within the NIH guidelines, to a point where molecular biology could go on its own merry way, more or less. Is there any prospectus as to how long it may take, and what will be required to begin that loosening up with respect to environmental release, and with respect to human gene therapy.

Answer: Well, I think I share what Bob said about this matter of complexity, and the complexity of the environmental problem being greater than the problem of gene therapy. It is an enormously complex problem. And therefore, I would not be optimistic that the time frame that you referred to would be applicable there. I think that until the problem is solved of how to get meaningful data about environmental release, without environmental release, that is, in some kind of confined arrangement, which is essentially what was put in place by   was the establishment of the confinement requirements of NIH. Until something of that kind is found, and I haven't seen anyone suggesting yet as to how that would be done, the rate at which information is going to grow and the consequences of environmental release is going to be low.

Question: As to your proposed structure, I think I have heard it said that the United States primary exports today are both medicine and agriculture. And both of these areas are impacted by regulation. My question is, if we have a tendency to overregulate here, compared to other countries, what is our, what is going to happen to our position in the world in these areas? In other words, what is the international picture look like? What is going on in other countries?

Answer: Bohrer: Well, let me give a two part answer to that question. I think the first part is that the concern about what we do to our relative economic standing by our regulatory structure is a well founded one. It does have an impact. Actually, it is a three part answer. For one, it does have an impact that is a well-founded concern. Number two, I am very uncertain, myself, about what that impact should be in my own normative system. I am just not quite sure how I want to deal with the argument that well, we ought to do it because they are doing it in Germany, and they will get ahead of us. I am not convinced that is a valid argument. Number three, I think it is incumbent upon us, as a leader in the development of the technology, to develop international standards and accords on some of these issues.

FOOTNOTES

1. For an account of some of the debate at Asilomar, see Berg, Baltimore, Brenner, Robin, and Singer, Asilomar Conference on Recombinant DNA Molecules, 188 Science 991-94 (1975).

2. Guidelines for Research on Recombinant DNA Molecules, 41 Fed. Reg. 27902 (1976).

3. Id.

4. Stetten, Freedom of Inquiry, 81 Genetics 416 (1975).

5. See Krimsky, Local Monitoring of Biotechnology: The Second Wave of Recombinant DNA Laws, 5 Recombinant DNA Bulletin 79-84 (1982).

6. Ord. No. 955, Final Publication No. 2092 (April 27, 1981).

7. Frederickson's views were expressed at length in "The Public Governance of Science." 3 Man and Medicine 77 (1978).

8. Foundation on Economic Trends, et al., v. Heckler, 756 F.2d 143 (D.C. Cir. 1985).

9. Diamond v. Chakrabarty, 447 U.S. 303 (1980).

10. See "Guidelines for Research Involving Recombinant DNA Molecules," 49 Fed. Reg. 46266 (1984).

11.   See note 7, supra.

12.   Yoxen, The Gene Business, (1983).

13.   Although the original text is unfootnoted, see Tanaka, The Japanese Legal System, chap. 4 (1976), for a more thorough account of the Japanese bar.

14.   For a very similar view to that expressed in the editorial cited in the text written for a more sophisticated audience, see Huber, Safety and the Second Best, the Hazards of Public Risk Management in the Courts, 85 Col. L. Rev. 277 (1985).

15.   Hybritech, Inc. v. Monoclonal Antibodies, Inc., 623 F.Supp. 1344 (1985).

16.   Pharmacia v. Hybritech, Inc., 224 U.S.P.Q. 975 (1985).

17.   OrthoDiagnostic Systems, Inc., v. Becton Dickinson Monoclonal Center, Inc., No. 84-519 (D. Del. complaint filed 9/21/84).

18.   For a more recent account of the stock market's infatuation with Genentech, see "Biotech's First Superstar," Business Week, April 14, 1986, at p.68.

19.   See Diamond v. Chakrabarty, supra note 9.

20.   Securities and Exchange Commission, 15 U.S.C. sec. 1 et seq.

21.   15 U.S.C. sec. 77-78.

22.   Accord, Huber, supra note 14.

23.   29 U.S.C. sec. 651 et seq.

24.   See Export Administration Act, 540 U.S.C. sec. 2401 et seq.

25.   Federal Noxious Weed Act of 1974, 7 U.S.C. sec. 2801-2813.

26.   See also Toxic Substances Control Act, 15 U.S.C. sec. 2601-2929, Federal Plant Pest Act, 7 U.S.C. 150 et seq., 7 U.S.C. sec. 166-177.

27. See Guidelines for Research on Recombinant DNA Molecules, 48 Fed. Reg. 24580 (1983).

28. See "Recombinant DNA Research, Request for Public Comment on Points to Consider in the Design and Submission of Human Somatic-Cell Gene Therapy Protocols," 50 Fed. Reg. 2940 (1985).

29. Plant Variety Protection Act, 35 U.S.C. sec. 101.

30. 447 U.S. 303 (1980).

31. See Pinon's discussion of the 1973 discovery of Cohen and Boyer, supra at 5.

32. The license agreement referred to is reproduced in Areen, King, Goldberg, and Capron, Law, Science and Medicine, (1984) at 109-113.

33. For an interesting follow-up to this point and to the Cohen-Boyer patent licensed by Stanford, see "Cetus Corporation Drops License from Stanford Under Cohen-Boyer Patents," 4 Biotech L. Report 130 (1985).

34. Id.

35. Roche Products, Inc., v. Bolar Pharmaceutical Co., Inc., 733 F.2d 858.

36. Supra at note 30.

37. See 21 U.S.C. sec. 510(k).

38. See "Proposal for a Coordinated Framework for Regulation of Biotechnology," 49 Fed. Reg. 50856 (1984).

39. The FDA clarified its official position in the Proposal cited id.

40. Recombinant DNA Committee, Office of Biologics Research and Review, Food and Drug Administration, April 10, 1985.

41. 35 U.S.C. sec. 156 (1984).

42. Convention of Budapest (Budapest Treaty of 1977).

43. E.g., Cal. Corp. Code sec. 25000 et seq.

44. See note 38, supra.

45. See Foundation on Economic Trends v. Heckler, 756 F.2d 143 (D.C. Cir. 1985).

46. See 49 Fed. Reg. 36052-054 (1984), where the RAC considered a request from Jeremy Rifkin of the Foundation on Economic Trends that the RAC "postpone its consideration of the Shiga-like toxin experiment . . . until an adequate mechanism is developed . . . to determine the potential military applications of this and certain other types of recombinant DNA research."

47. See note 28 supra. Dr. Grobstein was a member of the working group that produced the "Points to Consider."

48. Perhaps the best general source of information on the relative state of the industry around the world is Commercial Biotechnology: An International Analysis, (U.S. Govt. 1984).

49. Wickes was one of the more widely publicized San Diego companies to file for a chapter 11 federal bankruptcy reorganization in the early 80s.

50. See Foundation on Economic Trends v. Heckler, 756 F.2d 143 (D.C. Cir. 1985).

51. Compare the "Guidelines" cited in note 2 with those cited in note 8, supra.

52. See Diamond v. Chakrabarty, 447 U.S. 303 (1980).

53. No claim is made that it is an original idea to link any of these factors to risk perception. Several of these factors, in one form or another, were explored as a determinant of risk attitudes by Slovic, Fischhoff and Lichtenstein, "Facts and Fears: Understanding Perceived Risk," in Societal Risk Assessment, (Schwing and Albers, Jr., eds. 1980). Peter Huber has written extensively about the new/old factor, most recently in Safety and the Second Best, cited in note 14, supra. Slovic, et al., have not attempted to extrapolate from their excellent work on risk

perception, to the level of legislative or regulatory decision-making. Huber, on the other hand, quite clearly feels that any interjection of subjective risk perception into the decision-making process is an unmitigated evil. I think that the model I describe here is not merely a positivist description of social decision-making under conditions of uncertainty, but is also normative. I think it is both necessary and proper to allow our subjective judgments about relative risks and relative values to impact on our political choices.

54. See note 28, supra.

55. See Friedmann's description of proposed protocols for human somatic cell therapy, supra.

56. Mr. Rifkin is the founder of the Foundation on Economic Trends, which was the plaintiff in the case cited in aote 7, supra, and which is the leading organization to oppose expanded uses of R-DNA.

57. The EPA has recognized this problem, to a degree, in their initial pronouncements under their pesticide review authority, see 49 Fed. Reg. 50886 (1984).

58. See Slovic, et al., note 52, supra.

59. See Huber, supra, note 51.

60. The relationship between decision-making under conditions of uncertainty and the burden of proof is discussed at length in Bohrer, Fear and Trembling in the Twentieth Century: Technological Risk, Uncertainty, and Emotional Distress, 1984 Wisc. L. Rev 83 (1984).

61. See "Proposal for a Coordinated Framework for Regulation of Biotechnology," 49 Fed., e.g., 50856 (1984).

62. Vermont Yankee Nuclear Power Corp. v. Natural Resources Defense Council, Inc., 435 U.S. 519 (1978).

63. See "Report of the President's Commission on the Accident at Three Mile Island," (1979).

64. 51 Fed.Reg. 23325 (June 26, 1986).

65. See 51 Fed.Reg. 23326-23335.

ABOUT THE AUTHORS

ROBERT A. BOHRER

Mr. Bohrer is Professor of Law at California Western School of Law, where he has introduced the course "The Regulation of Safety." Professor Bohrer received a B.A. from Haverford College, his J.D. from the University of Illinois, and an LL.M. from Harvard Law School. Among his publications is "Fear and Trembling in the Twentieth Century: Technological Risk, Uncertainty and Emotional Distress," which appeared in the 1984 Wisconsin Law Review.

HARRY L. CASARI

Mr. Casari is an audit partner with the San Diego office of Arthur Young & Company. He is in charge of the office's Emerging Companies High Technology Group. His specialties include assisting companies with capital formation and assisting their management with related problems.

VINCENT FRANK

Mr. Frank is Executive Vice-President and General Counsel of Molecular Biosystems, Inc., a publicly-held biotechnology company, and as such acts as its chief administrative, financial, and legal officer. After receiving his B.A. from the University of Rochester and his J.D. from Northwestern, he practiced securities, corporate finance, and venture capital law as a partner in a Chicago law firm before helping to start Molecular Biosystems, Inc.

DR. THEODORE FRIEDMANN

Dr. Friedmann received his M.D. at the University of Pennsylvania, and was a Postdoctoral Fellow at the Salk Institute. He is a Professor in the Department of Pediatrics at the U.C.S.D. School of Medicine and is currently studying the molecular biology of human genetic disease and the development of model systems for studying the feasibility of gene therapy.

DR. CLIFFORD GROBSTEIN

Dr. Grobstein is currently Professor of Biological Science and Public Policy at U.C.S.D. During his distinguished career he has also been, among other things, Senior Research Fellow at the National Cancer Institute, Professor of Biology at Stanford, Dean of U.C.S.D.'s School of Medicine, and Vice Chancellor of Health Sciences at U.C.S.D. He has also been actively involved with the National Academy of Sciences, the National Institutes of Health, and the National Science Foundation, among other organizations, and has served on the President's Science Advisory Council. Dr. Grobstein has written widely on topics ranging from bio-chemistry to the ethical problems of in-vitro fertilization.

NED ISRAELSEN

Following an outstanding undergraduate science career at the University of Utah (magna cum laude in chemistry and biochemistry), Mr. Israelsen received his J.D. from George Washington University Law Center (Order of the Coif, Law Review). He now draws upon his knowledge of science as a partner in the San Diego firm of Knobbe, Martens, Olson & Bear, where he specializes in patent and intellectual property law, primarily in the bio-technology and pharmaceutical fields.

RAY W. McKEWON

Mr. McKewon, of the venture capital firm of McKewon & Timmons, was also a founder and managing partner of Enterprise Management Company ("EMC"), a venture capital firm, and has a great deal of experience in the bio-med, high-tech field. Mr. McKewon has arranged venture capital for Immunetech, Inc., Impulse Enterprise, and Cytotech, Inc. He was the first President of both Immunetech and Cytotech, and currently serves as Board Chairman for both companies. He also continues to serve as a member of the board of Impulse Enterprise. Mr. McKewon, who has an M.B.A. from Pepperdine University, has also been an account executive for Bache & Co. and raised capital for ENI Exploration Company.

DR. RAMON PINON

Dr. Pinon is Associate Professor in the Department of Biology at U.C.S.D. After receiving his Ph.D in Physics from Brown University, he focused his research in the field of genetics as a Postdoctoral Fellow at the University of Washington from 1965-1972. Dr. Pinon was a member of the NIH-RAC (Recombinant-DNA Advisory Committee) Panel which was responsible for the review and revision of the NIH Guidelines for R-DNA research from 1979-1982.

WILLIAM L. RESPESS

Mr. Respess is Vice-President and General Counsel of Gen-Probe, Inc., and was previously Vice-President and General Counsel of Hybritech, Inc. Mr. Respess received his Ph.D. in organic chemistry at M.I.T., and his J.D. at George Washington University. Prior to joining Hybritech he was a partner at the Los Angeles law firm of Lyon & Lyon, specializing in patent litigation in the federal courts.